T0140434

Springer Theses

Recognizing Outstanding Ph.D. Research

More information about this series at http://www.springer.com/series/8790

Aims and Scope

The series "Springer Theses" brings together a selection of the very best Ph.D. theses from around the world and across the physical sciences. Nominated and endorsed by two recognized specialists, each published volume has been selected for its scientific excellence and the high impact of its contents for the pertinent field of research. For greater accessibility to non-specialists, the published versions include an extended introduction, as well as a foreword by the student's supervisor explaining the special relevance of the work for the field. As a whole, the series will provide a valuable resource both for newcomers to the research fields described, and for other scientists seeking detailed background information on special questions. Finally, it provides an accredited documentation of the valuable contributions made by today's younger generation of scientists.

Theses are accepted into the series by invited nomination only and must fulfill all of the following criteria

- They must be written in good English.
- The topic should fall within the confines of Chemistry, Physics, Earth Sciences, Engineering and related interdisciplinary fields such as Materials, Nanoscience, Chemical Engineering, Complex Systems and Biophysics.
- The work reported in the thesis must represent a significant scientific advance.
- If the thesis includes previously published material, permission to reproduce this must be gained from the respective copyright holder.
- They must have been examined and passed during the 12 months prior to nomination.
- Each thesis should include a foreword by the supervisor outlining the significance of its content.
- The theses should have a clearly defined structure including an introduction accessible to scientists not expert in that particular field.

Li Yi

Study of Quark Gluon Plasma By Particle Correlations in Heavy Ion Collisions

Doctoral Thesis accepted by Purdue University,
West Lafayette, Indiana, USA

 Springer

Li Yi
Wright Laboratory
Yale University
New Haven, Connecticut, USA

ISSN 2190-5053   ISSN 2190-5061   (electronic)
Springer Theses
ISBN 978-1-4939-8215-8   ISBN 978-1-4939-6487-1   (eBook)
DOI 10.1007/978-1-4939-6487-1

Printed on acid-free paper

This Springer imprint is published by Springer Nature
The registered company is Springer Science+Business Media LLC New York

To my parents.

Supervisor's Foreword

Li Yi is a rare case. She is the kind of student every professor wants to have. She came to Purdue University in 2010. She quickly plugged into my research group upon her arrival and finished her Ph.D. in four and a half years. As you will see, her thesis consisted of three major parts. They resulted in four peer-reviewed journal publications: one *Physical Review* article (Chap. 3) and three *Physics Letters* (Chaps. 4 and 5). Chapter 3 describes the measurement of the third-order harmonic flow using two-particle correlations; Chap. 4 deals with the isolation of flow and non-flow contributions to particle correlations in gold–gold collisions; and Chap. 5 investigates the long-range longitudinal correlations in small system of deuteron–gold collisions. The first two are related to the hydrodynamic transport properties of the quark-gluon plasma created in gold–gold collisions. The last pertains to the question whether hydrodynamics is applicable to small systems, such as deuteron–gold collisions, and whether the quark-gluon plasma can be formed in those small-system collisions.

The thesis was conducted with the STAR experiment at the Relativistic Heavy-Ion Collider at Brookhaven National Laboratory, where the center-of-mass energy of the collision system was a factor of 100 larger than the rest mass of the colliding nuclei. The research in this thesis is highly relevant to our quest for deeper understanding of the fundamental theory of nature—the quantum chromodynamics. In particular, the results obtained in Chap. 5 challenge the interpretation of previous works from several other experiments on small systems and provoke a fresh look at the physics of hydrodynamics and particle correlations pertinent to high-energy nuclear collisions. The interpretation is unsettled yet and still heatedly debated. I believe Li Yi's work and interpretation will sustain the time test.

The thesis is full of gems. You will enjoy reading it.

Professor of Physics   Fuqiang Wang

Acknowledgments

On the journey to my doctoral degree, I have been blessed by many individuals, totaling more than what can be listed here.

First and foremost, I would like to thank my advisor Prof. Fuqiang Wang for his guidance and support over the years on my research and my career. Prof. Wang guided me through the adventure of heavy-ion physics with his inspiring ideas, patience in mentoring, prudent attitude toward data analysis, and critical feedback on the results. I also would like to thank Prof. Wei Xie, Prof. Andrew Hirsch, Prof. Rolf Scharenberg, and Prof. Brijish Srivastava for their helpful advices and discussions during our weekly group meetings. I enjoyed and would like to express thanks to the valuable and fun discussions with Prof. Denes Molnar through his heavy-ion course and in regard to my thesis. I also would like to thank Prof. Matthew L. Lister for serving on my thesis committee.

I would like to express my thanks to my fellow graduate students in high-energy heavy-ion group for the daily discussions: Tyler Browning, Liang He, Xin Li, Kun Jiang, Kurt Jung, Michael Skoby, Deke Sun, Jian Sun, and Quan Wang. I received much encouragement and help from Joshua Konzer from the beginning of my studies at Purdue through his graduation. I feel grateful to David Garand for providing numerous feedback on my presentations and paper writings. I would like to thank Mustafa Mustafa for teaching me the STAR software and sharing his interesting and inspiring stories. I also would like to give my special thanks to our group member Lingshan Xu and non-group member Yanzhu Ji for their friendship and joyful memories together over the years.

I am thankful to the STAR collaborators who helped me with my thesis work and who taught me a great deal in heavy-ion physics: the jetlike correlation PWG convenor Saskia Mioduszewski, the bulk correlation group convenor and nonflow paper GPC chair Hiroshi Masui, the $d+$Au ridge paper GPC chair Nu Xu, and members Helen Caines and Daniel McDonald. I also would like to thank Zhangbu Xu for introducing me to the heavy-ion field when I was in college.

Finally I would like to thank my parents for their understanding and love.

Contents

Abbreviations

ADC	Analog-to-Digital Converter
BBC	Beam-Beam Counter
BNL	Brookhaven National Laboratory
CTB	Central Trigger Barrel
dca	Distances of closest approach
FTPC	Forward Time Projection Chamber
LHC	Large Hadron Collider
MB	Minimum bias trigger data
QCD	Quantum Chromodynamics
QGP	Quark-Gluon Plasma
RP	Reaction plane
RHIC	Relativistic Heavy-Ion Collider
STAR	Solenoid Tracker at RHIC
TPC	Time Projection Chamber
ZDC	Zero Degree Calorimeter

Abstract

Yi, Li, Ph.D., Purdue University, December 2014. Study Quark-Gluon Plasma by Particle Correlations in Heavy-Ion Collisions. Major Professor: Fuqiang Wang.

A strongly interacting quark-gluon plasma (QGP) is believed to be created in relativistic heavy-ion collisions at the Relativistic Heavy-Ion Collider (RHIC). One of the important tools to study the properties of the QGP is the two-particle (dihadron) angular correlations. In dihadron correlations, the two major contributions are jet correlations and anisotropic collective flows of the QGP. While jet correlations probe jet-QGP interactions, anisotropic flows provide information about the thermodynamic properties of the QGP. Particularly, the third harmonic flow (v_3) is not only sensitive to the ratio of the QGP shear viscosity to entropy density but also the initial energy density fluctuations. This thesis provides the first v_3 measurement in Au + Au collisions at $\sqrt{s_{NN}} = 200\,\mathrm{GeV}$ from the STAR experiment. The $\Delta\eta$-gap, multiplicity, and p_T dependence of the v_3 are reported along with the comparisons with hydrodynamic calculations.

In heavy-ion collisions, the two-particle cumulant flow measurement is contaminated by nonflow correlations, such as contributions from jet correlations. An accurate flow measurement is crucial for the determination of the QGP shear viscosity to entropy density ratio. This thesis provides a data-driven method to isolate the $\Delta\eta$-dependent and $\Delta\eta$-independent components in the two-particle cumulant measurement. The $\Delta\eta$-dependent term is associated with nonflow, while the $\Delta\eta$-independent term is associated with flow and flow fluctuations. It is found that in 20–30 % centrality Au + Au collisions at $200\,\mathrm{GeV}$, the flow fluctuation is 34 % relative to flow, and the nonflow relative to flow square is 5 % with $\Delta\eta$-gap > 0.7 for particles with $0.15 < p_T < 2\,\mathrm{GeV}/c$ in $|\eta| < 1$.

The recent observations of a long-range $\Delta\eta$ correlation (the ridge) in $p + p$ and $p + \mathrm{Pb}$ collisions at the Large Hadron Collider (LHC) raised the question of collective flow in these small systems, which has been traditionally considered control experiments for heavy-ion collision studies. This thesis provides a careful analysis of short- and long-range two-particle correlations in $d + \mathrm{Au}$ collisions at $200\,\mathrm{GeV}$ from the STAR experiment. The event activity selection affects the jetlike correlated yield in $d + \mathrm{Au}$ collisions. Therefore, a simple difference between high- and low-activity collisions cannot be readily interpreted as nonjet, anisotropic flow correlations. This thesis reports the near-side ridge yield as a function of $\Delta\eta$ and its ratio to the away-side jet-dominated correlated yield, as well as the ratio to the underlying event multiplicity. This thesis also analyzes the dihadron azimuthal correlations in terms of Fourier coefficients V_n. The V_2 is found to be independent of event multiplicity and is similar between Au-going and deuteron-going forward/backward rapidities. These dihadron correlation measurements in $d+$Au collisions should provide insights into the theoretical understanding of the physics mechanism for the near-side ridge in $d+$Au system and the possibility of collective flow and QGP formation in the small systems.

Chapter 1
Introduction

The fundamental constituents of nuclear matter are quarks and gluons, together called partons. Gluons mediate strong force between quarks. Because strong force between quarks increases with the distance between their separation, quarks and gluons are confined within hadrons as color neutral objects [1–3]. Free quarks or gluons have never been observed. However, Quantum Chromodynamics (QCD) [4–7], the fundamental theory governing strong interaction, predicts that quarks and gluons can exist in a deconfined state, called Quark Gluon Plasma (QGP) [8–11]. The QGP is a plasma in which quarks and gluons can move in an extended volume without being restricted to the hadron size. Relativistic heavy-ion collisions are used to create and study such a QGP state in laboratory.

1.1 Quark Gluon Plasma

The attractive force between a quark–antiquark pair is roughly constant at large distances. The gluon binding potential between quark and antiquark is therefore proportional to their distance. The linear potential confines the quarks within the hadron size at zero temperature, since the energy needed to separate quarks increases linearly with their distance [12, 13]. If one tries to isolate a single quark out of a hadron, the gluon binding potential between the quark and the rest constituents of the hadron becomes energetic enough to create the quark–antiquark pair such that the gluon field will be separated into two regions. Each region will form a hadron itself, which again confines quarks. This process prevents the creation of an isolated, free quark.

However, deconfinement can occur at high nuclear densities. When nuclear matter density is high enough, the hadrons are compressed into one another. The quarks cannot identify their original partners in the hadron anymore, as they find a considerably large number of neighboring quarks in their former hadron radius

© Springer Science+Business Media New York 2016
L. Yi, *Study of Quark Gluon Plasma By Particle Correlations in Heavy Ion Collisions*, Springer Theses, DOI 10.1007/978-1-4939-6487-1_1

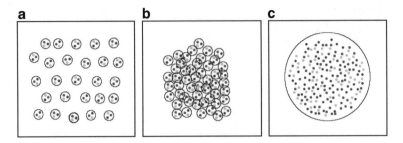

Fig. 1.1 With increasing nuclear matter density, matter changes from nuclear to quark matter. When nuclear matter density is low as shown in (**a**), quarks are confined inside their hadrons. As nuclear matter density increases, the hadrons are compressed and become close to each other as shown in (**b**). Eventually when the density is high enough, the hadron boundaries disappears and quarks are deconfined as shown in (**c**). Figure motivated by [14]

(see Fig. 1.1). When the distances between quarks and their new neighbors are short enough, the attractive forces on quarks become small. The quarks can therefore move over extended volume. Such change in quark motion is called deconfinement. Besides compressed into high nuclear density, deconfinement can also occur at high temperatures. At high temperature new quark–antiquark pairs can be produced, which effectively reduces the distances between quarks and thus makes the hadron boundaries disappear.

Another way to look at the transition from confinement to deconfinement is through its similarity to the Debye screening effect in electric plasmas. In dense plasma, each ion is surrounded by other ions and electrons. The effective Coulomb potential between an ion and its electron at certain distance away is screened by the surrounding ion and electron cloud with vanishing net charge. The effective potential decreases as charge density increases. When the effective potential is weak enough, the bound state of ion and electron becomes dissolved. The color screening in QGP is similar to the Debye screening in electric plasma by substituting electric charge with QCD color charge. However, the interaction properties of the force carrier for these two processes are different. While photons, the force carrier of electromagnetic interaction, do not interact between themselves, gluons, the force carrier of strong interaction, interact with each other. Gluons self-interactions make QCD binding energy $\propto r$ (r is the distance between charges), while the Coulomb potential is $\propto \frac{1}{r}$ at large distances.

1.2   Heavy-Ion Collisions

Ultra-relativistic heavy-ion collisions were proposed as a means to create QGP [15]. QGP exists at high temperatures ($> 170\,\text{MeV} \approx 2 \times 10^{12}\,\text{K}$ [16, 17]) or large baryon number densities (a few times of nuclear matter density). (Quarks and antiquarks have baryon numbers of $\frac{1}{3}$ and $-\frac{1}{3}$, respectively. The baryon number of a system is the sum of the baryon numbers of all its constituents.) However, normal nuclear matter exists at a comparatively low temperature (for example, even the center of the Sun is at merely 11×10^6 K $\approx 10^{-3}$ MeV) and low baryon number density (~ 0.17 per cubic fermi $= 0.17 \times 10^{-45}$ m^{-3}), QGP is not present in normal environments. There are generally two ways to achieve high energy density to form QGP. One way is to increase temperature to produce particle–antiparticle pairs without increasing net baryon number. The other way is to compress the system with many nucleons to increase baryon number density. Ultra-relativistic heavy-ion collisions create QGP primarily through the first one—increasing temperature. In ultra-relativistic heavy-ion collisions, two nuclei are accelerated close to the speed of light (99.995 % c) and are thus Lorentz contracted. When they collide with each other, the nuclei slow down through, naively speaking, multiple inelastic nucleon–nucleon collisions, depositing energy into the collision zone. If the energy density achieves the critical value (~ 1 GeV/fm^3 according to the QCD prediction [15]) for the phase transition, QGP will form. After a collision, high energy nucleons (baryonic matter) still have substantial forward/backward momentum to fly far away from the collision zone. Therefore, the net baryon number in the collision zone is small. However, the matter created has extremely high temperature. The collision process is illustrated as the curve in Fig. 1.2, starting from hadronic matter (nuclei), rising in temperature with low net baryon density into QGP phase, and then returning to hadronic phase in the end. In contrast, proton–proton ($p + p$) collisions at similar energies deposit less energy since there is only one nucleon–nucleon interaction.

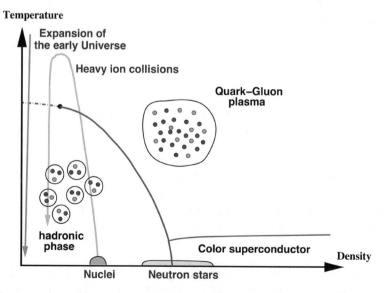

Fig. 1.2 A schematic view of nuclear matter phase diagram in QCD. Figure reprinted from [18]; copyrighted by Edmond Iancu

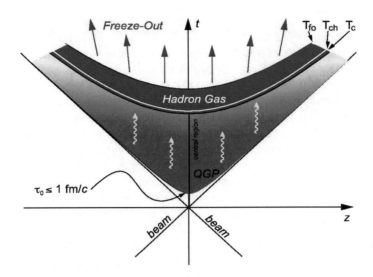

Fig. 1.3 Schematic view of QGP space-time evolution. T_c is the critical temperature for QGP to turn into hadrons. T_{ch} is the temperature when hadrons stop inelastic collisions and their hadron species become fixed. T_{fo} is the temperature when hadrons stop elastic collisions and free-stream to the detectors. Figure reprinted from [19] with permission of Springer

In a relativistic heavy-ion collision where QGP is formed, the system evolves through several space-time stages as depicted in Fig. 1.3. The inelastic nucleon-nucleon collision happens through parton–parton (quark or gluon) scattering. QGP is formed within $\sim 1\,\mathrm{fm}/c$ after the collision. The system begins to thermalize with further partonic scatterings. As the scatterings continue, the system expands in both longitudinal and transverse directions. The temperature decreases as the system expands. The photons and leptons radiated from the color medium leave the system without further (strong) interactions with QGP. When the temperature drops below the phase-transition critical value, the system starts to convert back into the hadronic state, to form baryons and mesons. The hadronization happens at $\sim 10\,\mathrm{fm}/c$. After hadronization, the system enters hadron gas phase. In the hadron gas phase, hadronic inelastic scatterings modify the particle species at the level of hadrons instead of partons. When further hadronic inelastic scattering ceases, hadron species are frozen. Particle elastic scatterings continue until their distances become too large as the system expands. Finally, elastic scattering ceases and particles stream freely and are recorded in detectors. The experimental observables could be charge, momentum, or energy of final state particles reconstructed in detectors. Final state particles carry the information about QGP as well as various stages of evolution.

The primary goal of high energy heavy-ion collision program is to create deconfined QGP and to investigate its properties, such as the critical temperature and the order of the phase transition, the equation of state, and the transport properties.

1.3 Collective Flow

The agreement between hydrodynamics calculation and experimental particle transverse momentum measurement indicates that, during QGP expansion, particles move collectively. This collectivity behavior is called flow phenomenon [20]. There are several forms of flow: the longitudinal flow, the axially symmetric radial flow, and the azimuthal anisotropic flow. The longitudinal flow will not be discussed in this thesis, while related literatures can be found in [21–23]. The radial flow and the anisotropic flow are in the transverse plane. Radial flow is driven by the QGP expansion in the radial direction. The amount of radial flow is generally governed by the cross sections of particle interactions (or in hydrodynamic language, viscosity). For particles of similar cross sections, a common radial flow velocity is customarily assumed. The heavier particles receive a larger boost in transverse momentum from the common radial flow velocity. The radial flow has been used to study kinetic freeze-out and the QGP equation of state [24, 25]. The azimuthal anisotropic flow is usually expanded in Fourier series for detailed study. The first harmonic anisotropic flow is called the dipole/directed flow. The second harmonic anisotropic flow is called the elliptic flow. The third anisotropic flow is called the triangular flow. The directed flow (or dipole flow) is the collective sidewards deflection of particles, as illustrated in Fig. 1.4. It was first observed in Ca + Ca and Nb + Nb collisions at 400 MeV/nucleon in the early 1980s [26]. The directed flow probes the pre-equilibrium and the thermalization stage as well as the initial-state fluctuation. The directed flow is small in the mid-rapidity region in high energy heavy-ion collisions. The references on directed flow can be found in [27–31]. This thesis focuses on the azimuthal anisotropic flow, particularly the elliptic and triangular harmonic flows. In the following context, the word "flow" refers to anisotropy flow.

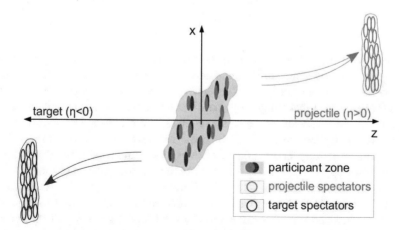

Fig. 1.4 Distribution of nucleons on the reaction plane. The collision axis is in the z direction. Spectators are referring to particles not participating in the collisions. The nuclear matter distribution in the participant collision zone has a sidewards deflection. Figure reprinted from [32]; copyrighted by the American Physical Society

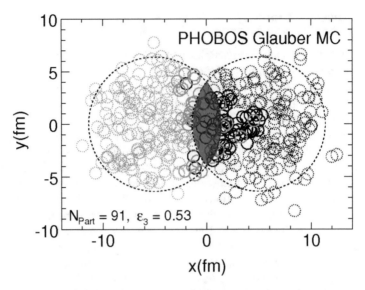

Fig. 1.5 Distribution of nucleons in the transverse plane. The *red shaded area* indicates smooth geometry overlap, while the *dark circles* indicate interacting nucleons (thereby defining the collision zone). Figure adapted with permission from [37]; copyrighted by the American Physical Society

In a semicentral collision, the pressure gradient is not uniform in azimuthal angle. Figure 1.5 shows the geometry of the overlap collision zone of two nuclei from the beam view. In spatial coordinates, the collision zone is almond shaped. The standard eccentricity of the overlap zone is defined by [33]

$$\epsilon_2 = \frac{\langle y^2 - x^2 \rangle}{\langle y^2 + x^2 \rangle}, \tag{1.1}$$

where (x, y) is the spatial position of the participant nucleons. The angle brackets are the average over all participant nucleons with unity weight. Other average definitions can be found in Refs. [34–36]. In Fig. 1.4, the pressure gradient along the x axis is larger than that along the y axis. Because the pressure gradient drives the direction of the expansion, as the system evolves, particles gain a larger momentum along the x axis than the y axis. The spatial anisotropy is thus transferred into a momentum anisotropy. The response of the final momentum anisotropy to the initial spatial anisotropy depends on the interaction strength among the constituents, which is related to the particle mean free path relative to the size of the collision system. When the mean free path is much larger than the size of the system, particles cannot interact, and therefore are unaware of the spatial geometry of the system. Hence, the particle momentum directions would be uniformly distributed as they are initially produced. When the mean free path is small relative to the system size, the information of the system's spatial distribution will be propagated to the particle momenta via interactions. The relative value of the mean free path to the system size is related to the shear viscosity. The smaller mean free path, the smaller shear viscosity. The comparison of the final anisotropy to the initial one, therefore, provides information about shear viscosity of the system. In addition to the overall geometry, there are event-by-event fluctuations (e.g., hot spots) in the collision zone. There exist higher order harmonics in the energy density distribution. The different order of coefficients respond differently to the system shear viscosity. Therefore, flow information can be used to constrain QGP viscosity calculations.

The particle momentum angular distribution can be written as a Fourier series:

$$\frac{dN}{d\phi} = \frac{N}{2\pi}[1 + 2v_1 \cos(\phi - \Psi_1) + 2v_2 \cos 2(\phi - \Psi_2) + 2v_3 \cos 3(\phi - \Psi_3) + \cdots]$$

$$(1.2)$$

$$= \frac{N}{2\pi}[1 + \sum_{n=1}^{\infty} 2v_n \cos n(\phi - \Psi_n)],$$

$$(1.3)$$

where ϕ is the particle azimuthal angle. The v_1 characterizes the directed flow; the v_2 characterizes the elliptic flow; the v_3 characterizes the triangular flow. They describe the magnitudes of particle momentum anisotropy. Ψ_n are the corresponding harmonic azimuthal angles. Of particular interest is the Ψ_2, called the second harmonic plane or second order of participant plane, which is determined by the initial participant nucleons (or partons) [38]. In Fig. 1.5, the short x axis is defined by the geometrical centers of the two colliding nuclei, and the z axis is the beam axis. The x-z plane is defined as the reaction plane. Due to fluctuations, Ψ_2 may not be the same as the reaction plane. The v_2 is the dominant term when the collision geometry is almond shaped (not perfectly overlapped as in Fig. 1.5). The v_3 is the third harmonic flow, which would be zero from the symmetry if the energy density in overlap region is smooth without fluctuations. Each v_n would have their own harmonic plane Ψ_n which could be different from each other.

Because the harmonic planes Ψ_n are not known a priori, v_n cannot be calculated directly from the single particle distribution as in Eq. (1.2). However, they can be obtained from two-particle correlations as follows. When single particle distribution follows Eq. (1.2), the two-particle distribution obeys

$$\frac{dN_{\text{pair}}}{d\Delta\eta} = \frac{N_{\text{pair}}}{2\pi}[1 + \sum_{n=1}^{\infty} 2V_n\{2\} \cos n\Delta\phi].$$

$$(1.4)$$

$V_n\{2\} = v_n^a \cdot v_n^b$ if there is no nonflow correlation (see discussion in Chap. 4). Here a, b stand for the two sets of particles used in the correlation measurements (where $\Delta\phi$ is the azimuthal opening angle between the particle pairs). When choosing a and b from the same kinematic region, v_n can be calculated by [39]

$$v_n\{2\} = \sqrt{\left\langle \frac{\sum_{i,j=1,i\neq j}^{M} \cos n(\phi_i - \phi_j)}{M(M-1)} \right\rangle}, \qquad (1.5)$$

where ϕ_i, ϕ_j are the azimuthal angles of particle pairs used for the correlation measurement; M is the number of particles used in each collision. Firstly, the average over M particles in each collision (one event) is taken. Then the average over all events is taken. (The weight used to average over all events can be the number of pairs or unity. If the number of pairs, $M(M-1)$, is used as the weight in the event average, then the calculation is equivalent to taking a single average over all pairs from all events.) The v_n calculated from two-particle correlations are called the two-particle cumulant flow. In a similar manner, one can also calculate v_n from four-particle correlations (see Chap. 4).

1.4 Jet–Medium Interactions

One way to probe QGP is to measure how jets are altered by QGP as they travel through it. A parton in a projectile nucleon interacts with a parton in a target nucleon. Occasionally large momentum transfers occur. A large momentum-transfer scattering is called hard scattering. In contrast, if the momentum transfer is small, the process is called a soft scattering. In a hard scattering, the large longitudinal energy is transferred into transverse plane. The final partons thus gain large transverse momenta. The final parton further fragments into a shower of partons. When partons travel through QGP medium, these partons exchange momentum and color with QGP, and thus are modified by QGP. These partons eventually hadronize into a cluster of hadrons, which are often called jets.

Jets as a probe for QGP have two primary advantages. Firstly, jet production can be calculated by perturbative QCD since it is produced in a large momentum transfer process. Secondly, jets are generated early in time. For a hard scattering of momentum transfer $Q_T = 2\,\text{GeV}/c$, the time scale of jet production is $\sim 1/Q_T \sim 0.1$ fm/c or less [40]. Jets have enough time to interact with QGP. By exploring how jets are modified, information about QGP medium can be gained.

On the other hand, jet as a probe is also complicated. Firstly, the partons themselves have complex time evolutions, even in the vacuum without QGP. They subsequently radiate gluons and can also split into quark–antiquark pairs. As a result, the nature of partons in a jet evolves as a function of time while branching into more partons. Secondly, QGP medium has a collective motion. The measured jet correlations therefore have a flow background. Thirdly, QGP medium is not static. The medium expands rapidly so that its temperature decreases. The interaction of jets with the dynamic medium is a challenging theoretical undertaking.

The first challenge above can be addressed by the following comparisons. Since the final measurement is the output of jet–medium interactions, it is necessary to have a baseline to understand the jet's behavior without the medium. $p + p$ collisions can be used as the vacuum baseline because QGP is generally not expected to form in $p + p$ collisions. Moreover, the parton distributions in the nucleon (except for low-x gluons), the partonic hard process cross sections, and the fragmentation of partons into hadrons are well understood [41]. Nonetheless, there is a defect in using $p + p$ collisions as the baseline. The initial multiple soft scattering effect in heavy-ion collisions, usually called Cronin effect[42], is not present in $p + p$ collisions. One way to include Cronin effect in the baseline is to study proton–nucleus (pA) or deuteron–nucleus (dA) collisions. In pA or dA collisions, Cronin effect is present, while QGP is generally not expected to form in such a small system. (However, there is a report of the possible formation of small droplets of QGP in p + Pb at the LHC energy. The search for QGP in $d + Au$ collisions at RHIC will be discussed in Chap. 5.)

The solution to the second challenge can be considered from two perspectives. From the experimental side, one can analyze the medium's collective motion, namely flow. By subtracting the flow, one will expect to obtain a clearer jet signal. From the theoretical point of view, a dynamic description of the jet–medium interactions should be incorporated. The third aforementioned challenge can also be addressed by realistic theoretical investigations. By comparing experimental measurements of jets and jetlike correlations to realistic and rigorous theoretical model studies, one hopes to learn valuable information about QGP.

1.5 Two-Particle $\Delta\eta$-$\Delta\phi$ Correlation and the Ridge

This section describes two-particle $\Delta\eta$-$\Delta\phi$ correlations and discusses the ridge structure and the away-side shoulder in heavy-ion collisions.

1.5.1 Why to Measure Two-Particle $\Delta\eta$–$\Delta\phi$ Correlation

Two-particle correlations have been used to study jets and jet–medium interactions. Ideally fully reconstructed jets tell a more complete story on how a jet interacts with the QGP medium than two-particle correlation studies. However, it is practically difficult to reconstruct a full jet in heavy-ion collisions, due to large number of particles. Jet is known to be a spray of particles which are the end products of one high energy parent parton. Instead of a parton spray, the experimental observation is the final hadrons reconstructed from the electrical signals. There are several challenges to reconstruct jets. Firstly, detectors have finite coverage and limited particle detection capabilities. Not all particles can be recorded in detectors. Secondly, the large number of final produced particles ($\sim$ thousands) in heavy-ion collisions makes it harder to distinguish which ones are from a jet, which are from the medium, and which are from the interaction between the jet and the medium. Thirdly, the different jet reconstruction algorithms generally lead to different jet results, as different jet algorithms may identify the jet with different particle constituents.

Because of these difficulties, high transverse momentum (p_T) particles are often used as a substitute for jets since they are usually the leading fragment of the jet. (At RHIC energy, a particle with p_T larger than $3\,\text{GeV}/c$ is generally considered to be a high-p_T particle.) The ratio of the high-p_T particle yield in heavy-ion collisions relative to that in $p + p$ or $d + \text{Au}$ baselines normalized by the number of binary collisions, called the nuclear modification factor, is found to be less than unity [43, 44]. This suppression of the single particle spectra at high p_T in heavy-ion collisions indicates jet energy loss in QGP. The suppression is often referred to as the jet-quenching phenomenon. However, the high-p_T single particle spectra method has two kinds of bias by design. Firstly, in order to have one high-p_T daughter particle, the selected high-p_T parent parton tends to give most of its energy to one single daughter particle in hard fragmentation. Such a bias selects the special collection of jet fragmentation for the study. Secondly, for jets in heavy-ion collisions, there is an additional bias the surface bias. The surface bias is illustrated in Fig. 1.6. The jet interacts with the medium and loses energy. The amount of energy loss depends on the path length of the jet through the medium. When triggered on high p_T, the jets with the shortest path length are more likely to be selected because they tend to lose less energy. These jets are mostly generated near the surface of the medium. As a result, they provide minimal information about the medium, since they interacted little with the medium. This surface bias limits the usefulness of these leading jets. On the other hand, due to momentum conservation, there is a recoil jet associated with the triggered high-p_T jet. While the jet with the triggered high-p_T particle has less interaction with the QGP, the recoil jet likely has maximal interaction with the QGP, because it has the longest path length to traverse

Fig. 1.6 The hard scattering producing the jets occurs near the edge of the medium. One of the jets (the trigger jet) leaves the medium soon after its formation and thus escapes without much further interaction with QGP, while the other one (the recoil jet) traverses the medium and is strongly modified by the medium. Figure reprinted from [18]; copyrighted by Edmond Iancu

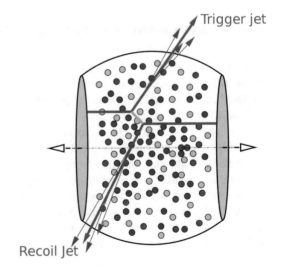

in the QGP. By studying the particles in the recoil jet, one can gain more information about jet–medium interactions. Two-particle correlations can be used to study the recoil jet associated with the trigger jet. The combination of high-p_T single particle spectra and two-particle correlations prove to be a powerful tool to shed light on how jets lose energy in QGP.

1.5.2 Two-Particle $\Delta\eta$–$\Delta\phi$ Correlation Method

Two-particle $\Delta\eta$–$\Delta\phi$ correlations measure the momentum angular distributions of the associated particles relative to the trigger particle. A trigger particle is usually defined as a particle with high p_T, which is likely from a jet. An associated particle is usually a lower p_T particle which may be from the same jet or the recoil jet, the medium or the jet–medium interaction. There are also cases when both the trigger and associated particles are low p_T particles. Low p_T two-particle correlation is sometimes referred to as untriggered two-particle correlation, which is often used to study medium properties. Two dimensions will be used in the following discussion: $\Delta\phi = \phi_{assoc} - \phi_{trig}$ is the azimuthal opening angle between the trigger and associated particles; $\Delta\eta = \eta_{assoc} - \eta_{trig}$ is the pseudo-rapidity separation between the particles. The region $\Delta\phi \approx 0$ is called the near side, where the associated particle azimuthal angle is similar to the trigger particle angle. The particles in the same jet as the trigger particle usually end up in the near side. The region $\Delta\phi \approx \pi$ is called the away side. The particles from the recoil jet will be on the away side. The $\Delta\eta$–$\Delta\phi$ two-particle (dihadron) correlation is given by

$$C(\Delta\eta, \Delta\phi) = \frac{1}{N_{trig}} \frac{d^2N}{d\Delta\eta d\Delta\phi} = \frac{1}{N_{trig}} \frac{S(\Delta\eta, \Delta\phi)/\epsilon_{assoc}}{B(\Delta\eta, \Delta\phi)/\langle B(\Delta\eta|_{100\%}, \Delta\phi)\rangle}. \tag{1.6}$$

Here $S = \frac{d^2N^{same}}{d\Delta\eta d\Delta\phi}$ is the raw dihadron correlation for pairs in the same event; and $B = \frac{d^2N^{mix}}{d\Delta\eta d\Delta\phi}$ is for trigger and associated particles from different events, which is called mixed event correction. $\langle B \rangle$ is the B average over $\Delta\phi$ at fixed $\Delta\eta|_{100\%}$. $\Delta\eta|_{100\%}$ is where the two particle acceptance is 100 %. The mixed event background serves as the correction for the detector's two-particle acceptance. For a 4π coverage detector with detection efficiency $\epsilon_{assoc} = 100\%$, $C(\Delta\eta, \Delta\phi) = \frac{S(\Delta\eta, \Delta\phi)}{N_{trig}}$. While the mixed event background does not contain the particle correlation existing in the same collision, it includes the detector acceptance information. Taking STAR experiment's Time Projection Chamber (TPC) as an example, there are 12 sectors in ϕ. The sector boundaries are the dead zones for particle detection. Particles across the sector boundaries have a lower probability to be reconstructed. Single particle deficiencies at sector boundaries affect two-particle correlations, especially when both particles cross the sector boundaries. Similarly, for the η direction and other detectors with non-uniform detector efficiencies, the single particle efficiency affects the two-particle correlations. Since the mixed event is also affected by the single particle detection efficiency, the raw dihadron correlation in the same event divided by the mixed event correlation can correct for the non-uniform detector efficiency. Meanwhile, detector has limited η acceptance. The STAR TPC has a good detection capacity in $-1 < \eta < 1$ for the single particle η acceptance, which gives a triangular shape for the dihadron correlations. The mixed event background also corrects the triangular shape in $\Delta\eta$. The mixed event dihadron correlation is normalized to be 100% at $|\Delta\eta|_{100\%} = 0$ when both the triggered and associated particles are in the TPC acceptance. The mixed events are also required to have primary vertices close to each other in the beam direction to resemble similar detector acceptance, and to have similar event characteristics, such as a similar number of particles. After divided by the mixed event background, the dihadron correlation in Eq. (1.6) is normalized by the total number of trigger particles used in the correlation study. The per trigger particle normalized dihadron correlation describes, on average, how many associated particles distributed in $\Delta\eta$–$\Delta\phi$ space for each trigger particle.

The dihadron correlation can be written as the sum of jetlike and flow correlations as shown in Eq. (1.7):

$$C(\Delta\eta, \Delta\phi) = J(\Delta\eta, \Delta\phi) + C_{\text{background}}(\Delta\eta, \Delta\phi); \tag{1.7}$$

$$C_{\text{background}}(\Delta\eta, \Delta\phi) = B_{ZYAM}(\Delta\eta)(1 + 2\sum \langle v_n^{\text{trigger}} v_n^{\text{associated}} \rangle \cos(n\Delta\phi)). \tag{1.8}$$

Here $C(\Delta\eta, \Delta\phi)$ is the measured dihadron correlation after the mixed event correction. $J(\Delta\eta, \Delta\phi)$ is the jetlike dihadron correlation. $C_{\text{background}}(\Delta\eta, \Delta\phi)$ is the flow modulated underlying event background. It can be expressed as Eq. (1.8) where v_n^{trigger} and $v_n^{\text{associated}}$ are the n-th harmonic azimuthal anisotropic flow for the trigger and associated particles, respectively. The azimuthal flow can be measured using several methods, such as the event-plane, the generating function, or the Q-cumulant method [39]. The difference between flow measurements from different methods is usually treated as systematic uncertainties. Normalization B_{ZYAM} is the uniform background value at each $\Delta\eta$, usually estimated by the Zero Yield At Minimum (ZYAM) method [45]. ZYAM assumes that the jet correlated yield is zero at its minimum:

$$J(\Delta\eta, \Delta\phi = \Delta\phi_{\text{min}}) = 0. \tag{1.9}$$

For various $\Delta\eta$, the jet yield minima $J(\Delta\eta, \Delta\phi = \Delta\phi_{\text{min}})$ can have different values and can also be located at different $\Delta\phi_{\text{min}}$.

The high-p_T dihadron $\Delta\eta$–$\Delta\phi$ correlation measurement shows a strongly suppressed away-side jet peak with a minimally modified near-side jet peak in central Au + Au collisions at $\sqrt{s_{NN}} = 200$ GeV [46, 47]. The similarity of near-side peaks in $p + p$, d + Au and in various centrality classes (see Sect. 2.2.5 for centrality definition) of Au + Au collisions, resembles the characteristics of hard scattering processes. The high-p_T particle away-side peaks are clearly observed in $p + p$, d + Au

and peripheral Au + Au collisions. In contrast, the high-p_T away-side peak in central
Au + Au collisions is remarkably diminished. This observation of the disappearance
of high-p_T away-side peak in central Au + Au collisions indicates that the recoil jet
loses significant energy when traversing QGP.

1.5.3 Near-Side Ridge

The ridge, a long-range azimuthal correlation, was first discovered in central
Au + Au collisions at $\sqrt{s_{NN}}$ = 200 GeV at the RHIC by the STAR experiment
[48]. The dihadron $\Delta\eta$–$\Delta\phi$ correlation was initially investigated with high-p_T
trigger particles for jet modification study. Before year 2010, the dominated flow
background, elliptic flow, was recognized and well studied. After the subtraction
of the elliptic flow background, a ridge structure was observed at small $|\Delta\phi|$ as a
near-side peak extended along large $\Delta\eta$, where jet contribution is minimal when
$|\Delta\eta| > 1$. As the top panels in Fig. 1.7 show [49], beside the near-side jet peak
at small $|\Delta\eta| < 1$, there is also a ridge on the near-side which is uniformly

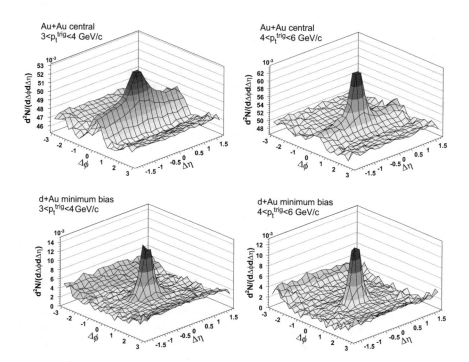

Fig. 1.7 Dihadron ($\Delta\eta$, $\Delta\phi$) correlations in d+Au (*lower panels*) and Au + Au (*higher panels*)
$\sqrt{s_{NN}}$ = 200 GeV per nucleon. Figure reprinted from [49]; copyrighted by the American Physical
Society

distributed in $\Delta\eta$ at large $|\Delta\eta| > 1$ in the STAR TPC acceptance $-2 < \Delta\eta < 2$ for both $3 < p_T^{\text{trigger}} < 4\,\text{GeV}/c$ and $4 < p_T^{\text{trigger}} < 6\,\text{GeV}/c$ in Au+Au central collisions. The ridge structure under the near-side jet peak is unknown. The naive expectation is that the ridge is also approximately uniform at $|\Delta\eta| < 1$. The ridge structure at even larger $\Delta\eta$ is reported by the PHOBOS experiment. Their detector $\Delta\eta$ coverage reached $-4 < \Delta\eta < 2$ for trigger particles $p_T^{\text{trigger}} > 2.5$ GeV/c and inclusive p_T for associated particles. The PHOBOS experiment found that the near-side ridge extends to $|\Delta\eta| \approx 4$ in central Au + Au collisions. The ridge yield is found to be largely independent of $\Delta\eta$ within their detectors acceptance. The collision multiplicity dependence of the ridge shows that the ridge yield decreases towards peripheral collisions, and the yield is zero when the collision has fewer than 100 participating nucleons for $p_T^{\text{trigger}} > 2.5\,\text{GeV}/c$ [50]. In contrast, there is no ridge in minimum bias d + Au collisions at the same collision energy $\sqrt{s_{\text{NN}}} = 200\,\text{GeV}$, as shown in the bottom panels of Fig. 1.7. There is also no ridge observed in minimum bias $p + p$ collisions at $\sqrt{s_{\text{NN}}} = 200\,\text{GeV}$ [48]. The near-side ridge discovered in central Au + Au collisions has three major features: long rapidity range, relatively small azimuthal angle $\Delta\phi$, and existence in events with a large number of particles.

Further studies, which will be discussed below, suggest that ridge physics has a soft origin (soft standing for small parton momentum transfer). For the two major contributions in dihadron correlations, jet and flow correlations: Flow correlations are the soft physics phenomenon; in contrast, jets are from the hard process. While the first ridge measurement was done with high-p_T trigger particles, further dihadron correlation measurements with low p_T particles showed that the ridge also exists for soft particles [51]. The confirmation of soft particle ridge suggests that jet physics may not be essential for the existence of near-side ridge. One should note, however, that soft near-side ridge measurements do not preclude a jet origin for the low p_T particles, since jets can also fragment into low p_T particles. Several measurements have been conducted to investigate whether the near-side ridge is originated from jet physics or not by comparing the behaviors of jet and ridge. Jet correlations from hard scattering have charge dependence. Jet dihadron correlations are stronger for pairs with opposite charges (unlike-sign) than pairs with the same charge (like-sign) due to charge conservation. The near-side ridge, however, is found to be the same for both unlike-sign and like-sign pairs so ridge behaves differently from jet dihadron correlations [51]. What's more, there is a strong correlation between the jet yield and the trigger particle p_T. The higher p_T^{trigger} particles come from more energetic jets. However, the near-side ridge has a weak dependence on the p_T^{trigger}. Additionally, the $p_T^{\text{associated}}$ spectra (particle yield distribution) of the near-side ridge is softer than the spectra from jets, but similar to inclusive particle production which is dominated by medium particles [52]. These measurements together suggest that the near-side ridge may not originate from jets, but rather is the feature of medium particles.

Among the medium effects, the azimuthal flow correlation is one possible candidate for the cause of near-side ridge. As discussed above, ridge is likely from medium particle correlations, so as the azimuthal flow. Besides, there are several similarities between ridge and flow behaviors. Firstly, similar to the near-side ridge,

Fig. 1.8 Glasma flux tubes
for the collision of two nuclei.
Figure reprinted from [53];
copyrighted by Annual
Reviews

the azimuthal flow is also more prominent in central collisions due to the larger multiplicity. Secondly, the large $\Delta\eta$ rapidity for near-side ridge requires an early time interaction. The early time interaction enables the possible connection between the two particles when they are separated at large $\Delta\eta$ in their final states. Using causality relationship, their interaction can be traced back to at or instantaneously after the encounter of the colliding nuclei[53]. Flow also develops at early times [54]. Thus azimuthal flow could be a viable explanation for the ridge. For the ridge measurement, the elliptic flow has already been removed in the dihadron ridge correlation. However, higher order harmonic azimuthal flows are not considered and they could exist in those ridge measurements.

Another explanation for the near-side ridge in heavy-ion collisions is the consequence of the early stage Glasma flux tube in concert with the later stage radial flow [53, 54]. The near-side ridge $\Delta\eta$ shape is not only long range, but also flat in $\Delta\eta$. The flatness in $\Delta\eta$ suggests that these correlations are independent of η. The Color Glass Condensate (CGC) theory [55, 56] predicts that the transverse color fields in the two nuclei transform into the so-called longitudinal Glasma flux tubes right after the collision, as Fig. 1.8 illustrates. Since the particles inside the same Glasma flux tube come from the same transverse position in the early times of the collision, the particles fragmented from the same tube share the same correlation regardless of their own rapidities. Thus, the particles produced from a Glasma flux tube resemble the long-range flatness in $\Delta\eta$ as the near-side ridge does. The small $|\Delta\phi|$ feature arises from the radial flow effect on the flux tube. The radial flow is the collective expansion of QGP medium in the transverse direction. The particles from the same flux tube experience the same radial flow velocity. The common radial flow collimates the outgoing particles into small $\Delta\phi$. The particles from the same Glasma flux tube are therefore focused by the radial flow to form the near-side ridge at small $\Delta\phi$ and be independent of $\Delta\eta$.

It is worthwhile to note that, even without radial flow, CGC predicts an enhanced two-gluon density at small $\Delta\phi$. However, such an effect is too small to explain the observed ridge in heavy-ion collisions. Such effect, on the other hand, may be related to the observed ridge in small system collisions at the LHC energies and perhaps at RHIC energies as well (see Chap. 5).

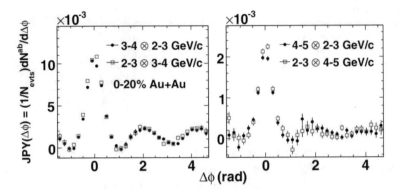

Fig. 1.9 The $\Delta\phi$ distributions of dihadron correlated per-trigger yield in 0–20 % Au + Au collisions at $\sqrt{s_{NN}}$ =200 GeV by the PHENIX experiment. Figure reprinted from [57]; copyrighted by the American Physical Society

1.5.4 Away-Side Shoulder

The away-side shoulder is a double-peak structure on the away side in central heavy-ion collisions [48]. Figure 1.9 shows the dihadron $\Delta\phi$ correlations in 0–20 % Au + Au collisions at $\sqrt{s_{NN}} = 200$ GeV by the PHENIX experiment [57]. The co-existence of near-side ridge and the away-side shoulder in central heavy-ion collisions may indicate the same origin.

The away-side shoulder has been studied as shown below. The per trigger away-side dihadron yield is found to increase towards central collisions, similar to the near-side ridge. Meanwhile, the average p_T of the particles on the away side associated with high-p_T trigger particles drops rapidly as the multiplicity increases, which means the particles in the shoulder become softer in central collisions [48]. The above two measurements show that the shoulder in central collisions has more and softer particles than in peripheral collisions. Moreover, the dihadron away-side yield depends on particle species. The baryon to meson ratio on the away side is enhanced in central collisions, which is quantitatively similar to that ratio for inclusive particles [57].

An earlier explanation for the away-side shoulder is the jet-medium effect. The supersonic jet shots through QGP medium, and mach-cone shock waves result in double-peak structure [58]. However, similar to near-side ridge measurement, the observation of the away-side shoulder is obtained from dihadron correlations with only elliptic flow subtracted. The initial state energy density fluctuation in spatial distribution can propagate into higher order final state particle momentum azimuthal correlations, such as the third harmonic v_3. Because the initial energy density distribution in the collision zone was previously considered to be smooth and the overlap region is perfectly almond shaped, higher order odd harmonics were not expected. The v_3 contribution was thus not subtracted in the previous dihadron correlation measurements. Glauber Monte Carlo simulations reveal the lumpiness in the initial state density which leads to the non-zero higher order harmonics in the final state, particularly the v_3 term [37]. The hydrodynamic flow v_3 is a soft physics, and it has a ridge shape on the near side and a double-peak shape on the away side. This particular shape of v_3 makes it the most natural explanation for the near-side ridge and away-side shoulder together.

The next question is now whether the hydrodynamics v_3 calculated with a fluctuating initial energy density can quantitatively describe the experimental measurements. Chapter 3 reports the v_3 measurement and its comparison with hydrodynamics.

References

1. M. Gell-Mann, A schematic model of baryons and mesons. Phys. Lett. **8**(3), 214–215 (1964). ISSN 0031-9163, http://dx.doi.org/10.1016/S0031-9163(64)92001-3. http://www.sciencedirect.com/science/article/pii/S0031916364920013
2. G. Zweig, An SU(3) Model for strong interaction symmetry and its breaking. Particle Phys. Phenomenol. Version 1 (1964). Preprint, http://cds.cern.ch/record/352337/
3. G. Zweig, An SU(3) model for strong interaction symmetry and its breaking. *Developments in the Quark Theory of Hadrons*, ed. by D. Lichtenberg, S. Rosen. Version 2 (Hadronic Press, Nonantum, 1980), pp. 22–101. http://cds.cern.ch/record/570209/
4. Y. Nambu, *A Systematics of Hadrons in Subnuclear Physics*. North-Holland, Amsterdam (1966), pp. 133–142
5. M.Y. Han, Y. Nambu, Three-triplet model with double SU(3) symmetry. Phys. Rev. **139**, B1006–B1010 (1965)
6. H.D. Politzer, Reliable perturbative results for strong interactions? Phys. Rev. Lett. **30**, 1346–1349 (1973)
7. D.J. Gross, F. Wilczek, Ultraviolet behavior of non-Abelian gauge theories. Phys. Rev. Lett. **30**, 1343–1346 (1973)
8. E.V. Shuryak, Quark-gluon plasma and hadronic production of leptons, photons and psions. Phys. Lett. B **78**, 150 (1978)
9. H. Bohr, H. Nielsen, Hadron production from a boiling quark soup: a thermodynamical quark model predicting particle ratios in hadronic collisions. Nucl. Phys. B **128**(2), 275–293 (1977)
10. M. Plümer, S. Raha, R.M. Weiner, How free is the quark-gluon-plasma? Nucl. Phys. A **418**, 549–557 (1984)
11. W.A. Zajc, The fluid nature of quark-gluon plasma Nucl. Phys. A **805**, 1–4, 283c–294c (2008); INPC 2007 Proceedings of the 23rd International Nuclear Physics Conference
12. J.F. Gunion, R.S. Willey, Hadronic spectroscopy for a linear quark containment potential. Phys. Rev. D, **12**, 174–186 (1975)
13. A. Chodos, R.L. Jaffe, K. Johnson, C.B. Thorn, V.F. Weisskopf, New extended model of hadrons. Phys. Rev. D. **9**, 3471–3495 (1974)
14. H. Satz, The analysis of dense matter. *Extreme States of Matter in Strong Interaction Physics*. Lecture Notes in Physics, vol. 841 (Springer, Berlin, 2012)
15. J. Bjorken, Highly relativistic nucleus-nucleus collisions: the central rapidity region. Phys. Rev. D **27**, 140 (1983)
16. J. Kuti, J. Polónyi, K. Szlachányi, Monte Carlo study of SU(2) gauge theory at finite temperature. Phys. Lett. B **98**(3), 199–204 (1981), ISSN 0370-2693, http://dx.doi.org/10.1016/0370-2693(81)90987-4. http://www.sciencedirect.com/science/article/pii/0370269381909874
17. L.D. McLerran, B. Svetitsky, A Monte Carlo study of SU(2) Yang-Mills theory at finite temperature. Phys. Lett. B **98**(3), 195–198 (1981)

18. E. Iancu, QCD in heavy ion collisions. arXiv:1205.0579 (2012)
19. M. Kliemant et al., Global properties of nucleus-nucleus collisions. *The Physics of the Quark-Gluon Plasma* (Springer, Berlin, 2010)
20. B. Muller, J.L. Nagle, Results from the relativistic heavy ion collider. Annu. Rev. Nucl. Part. Sci. **56**, 93–135 (2006)
21. J. Sollfrank, P. Huovinen, M. Kataja, P.V. Ruuskanen, M. Prakash, R. Venugopalan, Hydrodynamical description of 200A GeV/c S+Au collisions: hadron and electromagnetic spectra. Phys. Rev. C **55**, 392–410 (1997)
22. B. Schlei, U. Ornik, M. Plumer, D. Strottman, R. Weiner, Hydrodynamical analysis of single inclusive spectra and Bose-Einstein correlations for Pb + Pb at 160 A GeV. Phys. Lett. B **376**, 212–219 (1996)
23. The ATLAS collaboration, Measurement of forward-backward multiplicity correlations in lead-lead, proton-lead and proton-proton collisions with the ATLAS detector, in *25th International Conference on Ultra-relativistic Nucleus-nucleus Collisions, Kobe*. ATLAS-CONF-2015-051 (2015). https://cds.cern.ch/record/2055672
24. C. Hung, E.V. Shuryak, Equation of state, radial flow and freezeout in high-energy heavy ion collisions. Phys. Rev. C **57**, 1891–1906 (1998)
25. J. Adams et al., Identified particle distributions in pp and Au+Au collisions at $\sqrt{s_{NN}}$ = 200 GeV. Phys. Rev. Lett. **92**, 112301 (2004)
26. H.A. Gustafsson, H.H. Gutbrod, B. Kolb, H.Löhner, B. Ludewigt, A.M. Poskanzer, T. Renner, H. Riedesel, H.G. Ritter, A. Warwick, F. Weik, H. Wieman, Collective flow observed in relativistic nuclear collisions. Phys. Rev. Lett. **52**, 1590–1593 (1984)
27. J. Adams et al., Azimuthal anisotropy at RHIC: the first and fourth harmonics. Phys. Rev. Lett. **92**, 062301 (2004)
28. J. Adams et al., Directed flow in Au+Au collisions at $\sqrt{s_{NN}}$ = 62 GeV. Phys. Rev. C **73**, 034903 (2006)
29. B. Abelev et al., System-size independence of directed flow at the relativistic heavy-ion collider. Phys. Rev. Lett. **101**, 252301 (2008)
30. L. Adamczyk et al., Directed Flow of Identified Particles in Au + Au Collisions at $\sqrt{s_{NN}}$ = 200 GeV at RHIC. Phys. Rev. Lett. **108**, 202301 (2012)
31. L. Adamczyk et al., Beam-energy dependence of directed flow of protons, antiprotons and pions in Au+Au collisions. Phys. Rev. Lett. **112**, 162301 (2014)
32. B. Abelev et al., Directed flow of charged particles at midrapidity relative to the spectator plane in Pb-Pb collisions at $\sqrt{s_{NN}}$ = 2.76 TeV. Phys. Rev. Lett. **111**(23), 232302 (2013)
33. R.S. Bhalerao, J.-Y. Ollitrault, Eccentricity fluctuations and elliptic flow at RHIC. Phys. Lett. B **641**(3–4), 260–264 (2006)
34. D. Teaney, L. Yan, Nonlinearities in the harmonic spectrum of heavy ion collisions with ideal and viscous hydrodynamics. Phys. Rev. C **86**, 044908 (2012)
35. H. Niemi, G. Denicol, H. Holopainen, P. Huovinen, Event-by-event distributions of azimuthal asymmetries in ultrarelativistic heavy-ion collisions. Phys. Rev. C **87**, 054901 (2013)
36. F. Gardim, F. Grassi, M. Luzum, J.-Y. Ollitrault, Mapping the hydrodynamic response to the initial geometry in heavy-ion collisions. Phys. Rev. C **85**, 024908 (2012)
37. B. Alver, G. Roland, Collision-geometry fluctuations and triangular flow in heavy-ion collisions. Phys. Rev. C. **81**, 054905 (2010)
38. B. Alver et al., System size, energy, pseudorapidity, and centrality dependence of elliptic flow. Phys. Rev. Lett. **98**, 242302 (2007)
39. A. Bilandzic, R. Snellings, S. Voloshin, Flow analysis with cumulants: direct calculations. Phys. Rev. C **83**, 044913 (2011)
40. D. d'Enterria, B. Betz, High-p_T hadron suppression and jet quenching. *The Physics of the Quark-Gluon Plasma* (Springer, Berlin, 2010)
41. A.H. Mueller, *Advanced Series on Directions in High Energy Physics*, vol. 5 (World Scientific, Singapore, 1989)
42. J. Cronin et al., Production of hadrons at large transverse momentum at 200, 300, and 400 GeV. Phys. Rev. D **11**, 3105 (1975)

43. K. Adcox et al., Suppression of hadrons with large transverse momentum in central $Au + Au$ collisions at $\sqrt{s_{NN}} = 130\,\text{GeV}$. Phys. Rev. Lett. **88**, 022301 (2002)

44. C. Adler et al., Centrality dependence of high-p_T hadron suppression in Au + Au collisions at $\sqrt{s_{NN}} = 130\,\text{GeV}$," Phys. Rev. Lett. **89**, 202301 (2002)

45. N.N. Ajitanand, J.M. Alexander, P. Chung, W.G. Holzmann, M. Issah, R.A. Lacey, A. Shevel, A. Taranenko, P. Danielewicz, Decomposition of harmonic and jet contributions to particle-pair correlations at ultrarelativistic energies. Phys. Rev. C **72**, 011902 (2005)

46. C. Adler et al., Disappearance of back-to-back high p_T hadron correlations in central Au+Au collisions at $\sqrt{s_{NN}} = 200$ GeV. Phys. Rev. Lett. **90**, 082302 (2003)

47. J. Adams et al., Direct observation of dijets in central Au+Au collisions at $\sqrt{s_{NN}} = 200\,\text{GeV}$. Phys. Rev. Lett. **97**, 162301 (2006)

48. J. Adams et al., Distributions of charged hadrons associated with high transverse momentum particles in pp and Au + Au collisions at $\sqrt{s_{NN}} = 200\,\text{GeV}$. Phys. Rev. Lett. **95**, 152301 (2005)

49. B.I. Abelev et al., Long range rapidity correlations and jet production in high energy nuclear collisions. Phys. Rev. C **80**, 064912 (2009)

50. B. Alver et al., High transverse momentum triggered correlations over a large pseudorapidity acceptance in Au + Au collisions at $\sqrt{s_{NN}} = 200\,\text{GeV}$. Phys. Rev. Lett. **104**, 062301 (2010)

51. J. Adams et al., $\Delta\phi - \Delta\eta$ correlations in central Au+Au collisions at $\sqrt{s_{NN}} = 200\,\text{GeV}$. Phys. Rev. C **75**, 034901 (2007)

52. B.I. Abelev et al., Long range rapidity correlations and jet production in high energy nuclear collisions. Phys. Rev. C **80**, 064912 (2009)

53. J.J.-M.F. Gelis, E. Iancu, R. Venugopalan, The color glass condensate. Annu. Rev. Nucl. Part. Sci. **60**, 463–489 (2010)

54. S. Gavin, L. McLerran, G. Moschelli, Long range correlations and the soft ridge in relativistic nuclear collisions. Phys. Rev. C **79**, 051902 (2009)

55. L. McLerran, R. Venugopalan, Computing quark and gluon distribution functions for very large nuclei. Phys. Rev. D **49**, 2233–2241 (1994)

56. T. Lappi, L. McLerran, Some features of the glasma. Nucl. Phys. A **772**(3–4), 200–212 (2006)

57. A. Adare et al., Dihadron azimuthal correlations in Au+Au collisions at $\sqrt{s_{NN}} = 200\,\text{GeV}$. Phys. Rev. C **78**, 014901 (2008)

58. F. Wang, Supersonic jets in relativistic heavy-Řion collisions. AIP Conf. Proc. **892**(1), 417–420 (2007)

Chapter 2
STAR Experiment

This thesis work is performed with Au + Au and d + Au data recorded by the STAR detector at the RHIC accelerator.

2.1 Relativistic Heavy Ion Collider

The Relativistic Heavy Ion Collider (RHIC) is located at Brookhaven National Lab in Upton, New York on Long Island. RHIC is a versatile collider. It can accelerate various species of ions to a wide range in energy. The two major physics programs at RHIC are spin physics with polarized protons, and heavy-ion physics. Reviews on the spin physics program at RHIC can be found in [1, 2].

For heavy-ion physics, RHIC accelerates heavy nuclei of various species to collide at various energies. From its commissioning in 2000 to the present day (2014), RHIC has performed proton and proton (p + p), deuteron and gold (d + Au), copper and copper (Cu + Cu), gold and gold (Au + Au), copper and gold (Cu + Au), uranium and uranium (U + U), helium-3 and gold (He^3 + Au) collisions. RHIC has conducted a beam energy scan program for Au + Au collisions at the center of mass energy $\sqrt{s_{NN}}$ from 7.7 GeV to the top energy of 200 GeV per nucleon pair. The various energies facilitate the search for the possible critical point of QCD phase diagram [3–6].

As Fig. 2.1 shows, there are six collision points on RHIC's 3.8 km long storage ring, among which are four experiments. They are STAR at 6 o'clock, PHENIX at 8 o'clock, PHOBOS at 10 o'clock, and BRAHMS at 1 o'clock. While PHOBOS and BRAHMS finished their missions in 2005 and 2006, respectively, the STAR and PHENIX experiments are still operating as of 2015.

While serving as a heavy-ion collider, RHIC has been proposed as a possible facility to host a future electron-ion collider to study the partonic structure of nuclei [8, 9].

© Springer Science+Business Media New York 2016
L. Yi, *Study of Quark Gluon Plasma By Particle Correlations in Heavy Ion Collisions*, Springer Theses, DOI 10.1007/978-1-4939-6487-1_2

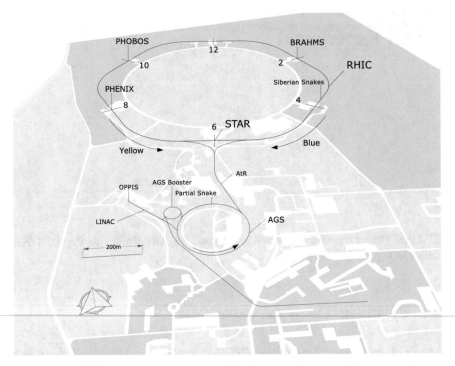

Fig. 2.1 The RHIC accelerator complex. Figure reprinted from [7]; copyrighted by Tai Sakuma

2.2 STAR Detector

The Solenoidal Tracker at RHIC (STAR) detector is a general purpose detector. STAR has various detector subsystems. The main detector component is the TPC. It has full azimuthal and approximately 2.5 units of rapidity coverage. Besides TPC, another two major detector subsystems used in this analysis are the Forward Time Projection Chamber (FTPC) and the Zero Degree Calorimeter (ZDC), as shown in Fig. 2.2.

The conventional coordinate system at STAR sets the Time Projection Chamber center as the origin point. The beam pipe direction is the z direction with the west direction as being positive. The x direction is pointing to the south and the y direction is pointing up. The x-y plane is the azimuthal angle plane. For the $d + Au$ collisions conducted in year 2003 and 2008, the deuteron beam moved towards the west, the positive z direction, and the gold beam moved towards the east, the negative z direction.

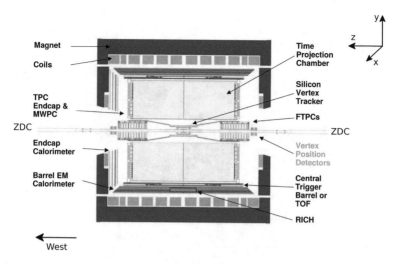

Fig. 2.2 An illustration of a cutaway side view of the STAR detector [10]. Figure modified from http://www.star.bnl.gov/public/trg/trouble/operating-trigger/introduction/star_detector. pdf; copyrighted by the STAR collaboration

2.2.1 Time Projection Chamber

The TPC is the primary tracking detector at STAR [11]. The TPC measures charged particle momentum and charge, and identifies particle species. It is 4 m in diameter and 4.2 m long, covering of $-1.2 < \eta < 1.2$ in pseudo-rapidity with high-quality tracking. With the STAR magnetic field of 0.5 T in the z direction, the TPC can measure particles with momentum greater than 150 MeV/c.

The TPC detects charged particles via their ionization in the TPC gas volume. As Fig. 2.3 illustrates, the thin conductive central membrane, the concentric field cage cylinders and two end caps provide a nearly uniform electric field along the beam pipe z direction in the TPC. When a charged particle traverses in the TPC gas, it ionizes the gas atoms and electrons are released from the atoms. The ionized electrons drift in the electric field until they enter the high field around the anode where the electrons avalanche. The anode is located at the Multi-Wire Proportional Chambers as the end cap readout. The amplitude of the current collected by wire is proportional to the ionization energy loss. The location of the wire portrays the position of the charged particle ionization in the x-y plane. Both east and west end caps have 12 sectors in ϕ with 45 pad rows along the radius. Each pad row could record an ionization location in x-y. Therefore, each particle track could have a maximum of 45 hit points of the recorded ionization locations. The particle reconstruction efficiency is low at the sector boundaries since there is no wire to collect current there, which means that the TPC has detecting deficiency at the sector boundaries. The z positions of the charged particle ionization locations are calculated by the product of electron drifting time and its drift velocity. The electron

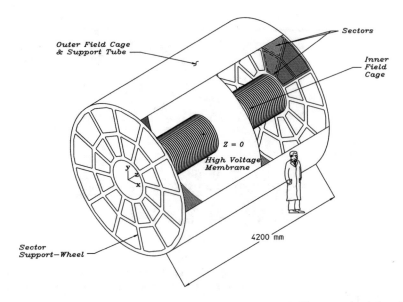

Fig. 2.3 The illustration of the Time Projection Chamber at STAR. Figure reprinted from [11] with permission of ELSEVIER; permission conveyed through Copyright Clearance Center, Inc.

drifting time is the time difference between the collision time given by other fast detectors and the time when the electrons reach the end cap to give an electronic signal. The TPC is filled with P10 gas (Ar 90 % + CH₄ 10 %) which provides a stable electron drift velocity that is insensitive to small variations of temperature and pressure. With the ionization points x, y, and z known along the particle path, the helix of the particle motion is reconstructed through track fitting algorithm. The particle helix and the STAR magnetic field magnitude together are used to determine the particle momentum and its charge sign from the Lorentz force equation of motion. The ionization energy loss dE/dx measured from the readout current is used to identify the particle species.

The particle tracks reconstructed from the TPC hit points are called global tracks. After the track reconstructions in the event are finished, the primary vertex of the collision, which is the estimate of the interaction point, is reconstructed from the global tracks. The closest distance from the primary vertex to the track helix is called the Distance of Closest Approach (*dca*). The tracks with *dca* less than 3 cm are refitted with the primary vertex included, which then become primary tracks. The primary track has better momentum resolution than the global track because the primary vertex position from all tracks is more precise than any other fit point on the single track if the track is from the primary vertex. However, when the track is from a secondary vertex, for example from a resonance decay, the primary track becomes less accurate than the global tracks. Hence, the 3 cm *dca* requirement is imposed for refitting of the primary tracks. To improve STAR's tracking ability in mid-rapidity, the Heavy Flavor Tracker (HFT) detector, which will not be discussed in detail in this thesis, was installed for data taking in 2014. HFT is composed of three silicon detectors with the inner layer installed right outside the beam pipe, therefore it has precise vertex determination ability. It significantly empowers the heavy flavor (charm or bottom quark) program in STAR where the precise decay vertex detection is crucial [12].

2.2.2 Forward Time Projection Chamber

Two cylindrical FTPCs measure charged particles with $2.5 < |\eta| < 4$ [13]. The FTPCs extend the STAR rapidity coverage. The major design difference of the FTPC from the TPC is that the electric drift field is in the radial direction instead of parallel to the beam axis. The radial field insures the high resolution at large particle density close to the beam pipe where the FTPCs operate.

The schematic diagram of an FTPC is shown in Fig. 2.4. In order to fit in the available space inside the TPC (see Fig. 2.2. The STAR TPC conceptual design was proposed at year 1993 [14]. The FTPC design was proposed at year 1998.), each FTPC is 75 cm in diameter by 120 cm long. The FTPCs share the common STAR magnetic field with the TPC, along the z direction. The radial electric field of the FTPC is provided by the high voltage inner electrode and the grounded outer cylinder wall. The FTPC readout system is located outside the cylinder chamber

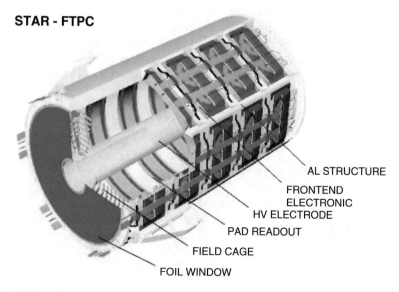

Fig. 2.4 The schematic diagram of the FTPC. Figure reprinted from [13] with permission of ELSEVIER; permission conveyed through Copyright Clearance Center, Inc.

surface for its radial drift field. The readout system has six sections in the azimuthal ϕ direction and 10 rows in z to afford a maximum of 10 hit points.

Similar to the TPC tracks, the FTPC tracks are reconstructed in two steps (global and then primary tracks) with one additional calibration procedure between these two steps (misalignment correction). The FTPC global tracks are reconstructed from the FTPC hits. The collision vertex reconstructed from the FTPC global tracks is called FTPC "pre-vertex." The primary vertex reconstructed from the TPC tracks has a better position resolution than the FTPC pre-vertex because TPC resolution is higher than the FTPC. The TPC and FTPC may potentially have a misalignment due to a possible shift (or rotation) of the FTPC mounting points relative to the TPC detector. Therefore, a misalignment correction procedure is performed based on the discrepancy between the FTPC pre-vertex and the TPC primary vertex. The FTPC primary tracks are reconstructed with the necessary corrections to align the FTPC tracking coordinate system with the TPC one. To summarize, the FTPC track reconstruction procedure is: reconstruction of FTPC global tracks, determination of FTPC pre-vertex, determination of the correction for misalignment between FTPC and TPC, and finally fitting the alignment corrected FTPC global tracks with TPC vertex to obtain the primary tracks. FTPC served in STAR data taking from year 2001 to year 2011.

2.2.3 *Zero Degree Calorimeter*

The ZDCs are two hadron calorimeters measuring the neutron energy along the beam pipe after the charged particles are bent away from the ZDC acceptance by the dipole magnets, as Fig. 2.5 shows [15]. Two ZDCs are located symmetrically at 18 m away from the collision intersection point on each side, with a horizontal acceptance of $\pm$ 5 cm. In relativistic heavy-ion collisions, the evaporation neutrons

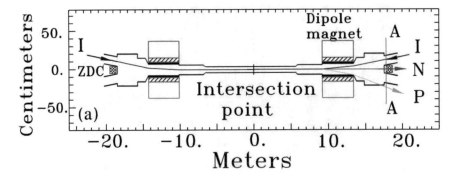

Fig. 2.5 The schematic diagram of the ZDC in the context of collisions. Figure reprinted from [15] with permission of Elsevier; permission conveyed through Copyright Clearance Center, Inc.

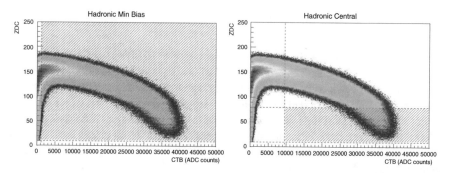

Fig. 2.6 An illustration of event characterization by the ZDC-measured neutral energy versus the CTB-measured charged particle multiplicity for minimum bias (*left*) and central triggered (*right*) Au + Au collisions

are emitted from spectators, which do not participate in the later collision. The ZDC measures the energy of those neutrons. Meanwhile, the east and west ZDC signal timing difference provides an estimate for the collision location.

The two main purposes of the ZDCs are event characterization and luminosity monitoring. Together with the Central Trigger Barrel (CTB), the ZDCs characterize the collision centrality. The CTB measures the charged particle multiplicity in $-1 < \eta < 1$ with full ϕ coverage. The CTB is arranged around the TPC and covers similar phase space as the TPC. The CTB records data faster than the TPC detector. The arch shaped color band in Fig. 2.6 (same for left and right panels) illustrates the ZDC measured neutral energy versus the CTB measured charged particle multiplicity distribution for Au + Au collisions. In peripheral Au + Au collisions, the gold ions are largely untouched and thus bent away by the dipole magnet. There are only a few neutron spectators. Hence, the ZDC signal is small. Meanwhile, the number of the produced charged particles from the collision is also small so that the CTB has a small signal. In semicentral Au + Au collisions, since the gold ion spectators are mostly shattered into nucleons, the number of neutron spectators reaches its maximum. The ZDC signal is thus at its maximum. The particle multiplicity in CTB for a semicentral collision lays between peripheral and central collision ones. A central collision has the larger mid-rapidity multiplicity compared to more peripheral one, because more nucleons are involved in a central collision. The central collision gives the maximum CTB signal. However, when all the nucleons participate in the interaction, the number of neutron spectators again becomes small, resulting in a small ZDC signal. The ZDC and CTB signals together can characterize the collision centrality of Au + Au collisions. CTB detector was later replaced by more advanced Time of Flight system (TOF) which has the similar coverage but can also extend STAR's particle identification dynamic range [16].

2.2.4 Minimum Bias Event

The conflict between high RHIC crossing rate ($\sim 10\,\text{MHz}$) and the low TPC response rate ($\sim 100\,\text{Hz}$) calls for the need of a trigger system to effectively select out interested events to record. The event characteristics are examined using the fast detectors. When an event passed a given trigger requirement, the trigger system sends a request to start the recording cycle for the slower but more precise detectors. Both the TPC and the FTPCs are slow detectors because the electrons need enough time to drift to the readout devices. Both the ZDCs and the CTB are fast detectors, and are used in the trigger system.

Various triggers are used to determine the interested events to record and the rest to discard. The left and right panels in Fig. 2.6, as already seen in Sect. 2.2.3, show an example of two kinds of triggers for Au + Au collisions: minimum bias (left panel) and central (right panel) trigger. The minimum bias (MB) triggered events are those in the red shaded area. While zero bias events are referred to all possible collisions, MB events are the best event sample estimation for the zero bias events with an efficient event recording. The typical trigger requirements for an MB event in heavy-ion collisions are

- there are ion bunches in both beam pipes for a possible collision,
- at least one neutron is detected in each ZDC detector so that there are interactions happening,
- the timing difference between two ZDC signals satisfies the correct window to assure the interaction point is inside the TPC, and
- at least 15 particles hit the CTB so that the interaction is an inelastic collision.

The central trigger is the subset of MB trigger with large number of particles hit the CTB. The central triggered events are most likely to have QGP formed. There are various triggers implemented for physics purposes and new detectors monitoring: high p_T jet events, beam polarization for spin studies, cosmic rays for calibrations, heavy flavor for quarkonium production, and so on. For d + Au collisions, the MB event is triggered by the ZDC in the gold-beam going direction only. The STAR trigger system is designed to allow that, in generally, the detector samples the maximum number of collisions provided by RHIC within the detector response time.

2.2.5 Centrality Definition in Heavy-Ion Collisions

The impact parameter of a collision, which is the distance between the centers of two colliding nuclei, varies from one event to another. A central collision with a zero impact parameter has a full overlap area, and tends to have the maximum energy density. A peripheral collision of two nuclei with large impact parameter has a small overlap zone, and tends to have lower energy density and be similar to a

$p+p$ collision. The collision systems with the various initial collision geometries are thus different. Experimentally, the setup of collision geometry cannot be controlled. On the other hand, in order to compare physical quantities in $A + A$ collisions (nucleus + nucleus) to the ones in $p + p$ collisions (nucleon + nucleon), one needs to know the number of equivalent binary nucleon + nucleon collisions (N_{coll}) in an $A + A$ collision. It is also natural to ask whether an observable is related to binary collisions N_{coll} or the number of participating nucleons N_{part}. Observables related to hard process usually depend on N_{coll}. Observables related to soft process usually depend on N_{part}. Neither N_{coll} nor N_{part} is experimentally accessible.

The Glauber Monte Carlo simulation has been used to model the collision geometry and links experimental observables with the theoretical b, N_{coll}, and N_{part} [17]. In high energy heavy-ion collisions, the de Broglie wavelengths of the high speed nucleons are smaller than their transverse size. Their total geometry cross section is approximately the sum of the individual nucleon + nucleon collision cross sections. The inputs to the Glauber simulation are the Wood–Saxon nuclear matter density and the inelastic nucleon-nucleon cross section. For a given b, the Glauber simulation generates the nucleon distribution in the nuclei according to the Wood–Saxon density profile. Further assumptions include that the nucleons move in straight lines, each nucleon + nucleon collision happens independently and the nucleons motion does not change after the nucleon + nucleon collisions. One nucleon in a nucleus can interact with several nucleons in the other nucleus if they are in its way. A nucleon + nucleon collision will happen if their distance is less than the nucleon + nucleon inelastic cross section. The values of N_{coll} and N_{part} are the output of the Glauber model for a given b. An observable needs to map the Glauber simulation to experimental data for a centrality definition. This observable should be a monotonic function of the impact parameter b. As Fig. 2.7 illustrates, the inclusive charged particle multiplicity can suit the purpose as a centrality definition observable: the larger the multiplicity, the smaller the b. The dashed lines are an illustration of the typical centrality binning. The illustrations of various collision geometries in the beam view are also depicted for several centrality classes.

While the centrality binning procedure seems straightforward, there are several factors one needs to keep in mind: the selection bias of the centrality observable, the fluctuations in the experimental observable and in the Glauber model parameters, and the detector acceptance effect. The centrality observable selection bias is a challenge especially for small systems due to the low particle multiplicity, as the case in $d + Au$ collisions (see Chap. 5). For example, events with an energetic jet tend to have a larger multiplicity than the average MB events. Selecting on high multiplicity would therefore bias events towards jet productions and bias jets towards larger energies. However, in heavy-ion collisions, the multiplicity is high and the particle production from jets is relatively unimportant compared to the underlying event multiplicity. Hence, the multiplicity selection bias is small in heavy-ion collisions.

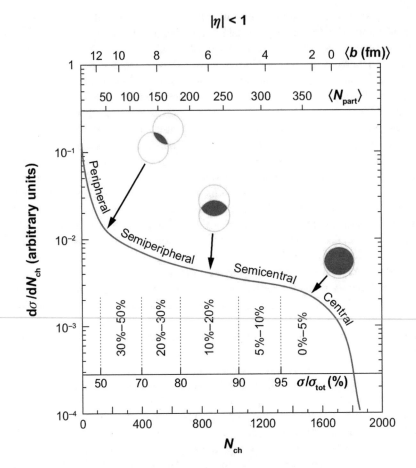

Fig. 2.7 An illustration of the correlation of the inclusive charged particle multiplicity and b and N_{part} from Glauber model for centrality definition in Au + Au collisions. Figure reprinted from [17]; copyrighted by Annual Reviews

References

1. E.C. Aschenauer et al., The RHIC spin program: achievements and future opportunities (2013)
2. W. Haeberli, Sources of polarized ions. Annu. Rev. Nucl. Part. Sci. **17**, 373–426 (1967)
3. M.A. Stephanov, Sign of kurtosis near the QCD critical point. Phys. Rev. Lett. **107**, 052301 (2011)
4. L. Adamczyk et al., Elliptic flow of identified hadrons in Au+Au collisions at $\sqrt{s_{NN}}$ = 7.7- 62.4 GeV. Phys. Rev. C **88**, 014902 (2013)
5. L. Adamczyk et al., Beam energy dependence of moments of the net-charge multiplicity distributions in Au + Au collisions at RHIC. Phys. Rev. Lett. **113**, 092301 (2014)
6. L. Adamczyk et al., Energy dependence of moments of net-proton multiplicity distributions at RHIC. Phys. Rev. Lett. **112**, 032302 (2014)

7. T. Sakuma, Inclusive jet and dijet production in polarized proton-proton collisions at 200 GeV at RHIC, ScD thesis, Massachusetts Institute of Technology, 2010
8. A. Deshpande et al., Study of the fundamental structure of matter with an electron ion collider. Annu. Rev. Nucl. Part. Sci. **55**, 165–228 (2005)
9. E.C. Aschenauer et al., eRHIC design study: an electron-ion collider at BNL. arXiv:1409.1633 (2014)
10. K. Ackermann et al., STAR detector overview. Nucl. Instrum. Methods **A499**(2–3), 624–632 (2003)
11. J. Anderson, A. Berkovitz, W. Betts, et al., The STAR time projection chamber: a unique tool for studying high multiplicity events at RHIC. Nucl. Instrum. Methods **A499**, 659–678 (2003)
12. H. Qiu, STAR heavy flavor tracker. Nucl. Phys. A **931**, 1141–1146 (2014)
13. K. Ackermann, F. Bieser, F. Brady et al., The forward time projection chamber in STAR. Nucl. Instrum. Methods **A499**(2–3), 713–719 (2003)
14. H. Wieman et al., STAR TPC at RHIC. IEEE Trans. Nucl. Sci. **44**, 671–678 (1997)
15. C. Adler, A. Denisov, E. Garcia, M. Murray, H. Stroebele, S. White, The RHIC zero degree calorimeters. Nucl. Instrum. Methods **A470**(3), 488–499 (2001)
16. P. Fachini et al., Proposal for a Large Area Time of Flight System for STAR (2004)
17. M.L. Miller, K. Reygers, S.J. Sanders, P. Steinberg, Glauber modeling in high-energy nuclear collisions. Ann. Rev. Nucl. Part. Sci. **57**, 205–243 (2007)

Chapter 3
Higher Harmonics v_3

This chapter of thesis work has been summarized in [4].

The particle momentum anisotropy distribution has been considered an important tool to study the QGP bulk properties. The particle momentum anisotropy distribution is usually analyzed in the form of the Fourier coefficients. In heavy ion physics, the first harmonic v_1, directed flow, and the second harmonic v_2, elliptic flow, have been extensively studied both experimentally and theoretically. The higher harmonic v_4 has also been studied and is primarily from the nonlinear medium response to the eccentricity ϵ_2 [Eq. (1.1)] of the initial elliptic shape. The odd higher harmonics have traditionally been considered to be zero assuming the initial energy density distribution is smooth and symmetric. However, a Monte Carlo simulation study [1] reported that event-by-event initial state geometry fluctuations can lead to nonzero higher order odd harmonics. By studying the higher harmonics, one could gain information about the lumpiness of the initial state energy density distribution. Each order of the anisotropy harmonics responds to the hydrodynamic viscosity differently. Therefore, the combination of the various harmonic measurements, for example v_2 and v_3 together, provides better constraints on the hydrodynamic viscosity than the traditional elliptic flow v_2 measurement alone. However, the higher harmonics v_n with $n > 3$ have nonlinear responses to the lower harmonics. Hence, for the purpose of extracting QGP bulk properties by comparing anisotropy measurements v_n with hydrodynamic calculations, v_3 is optimal.

3.1 Two-Particle Q-Cumulant Method

As discussed in Sect. 1.3, due to the unknown participant plane, the two-particle correlation method has been used in experiment to measure anisotropic flow. If using Eq. (1.5) directly, the nested loops over the particle pairs will become a CPU consuming process. The mathematical technique of cumulants is thus adopted to simplify the analysis calculation [2].

© Springer Science+Business Media New York 2016

L. Yi, *Study of Quark Gluon Plasma By Particle Correlations in Heavy Ion Collisions*, Springer Theses, DOI 10.1007/978-1-4939-6487-1_3

The cumulant is expressed in terms of the particle flow vector Q. For one event with M particles, the nth order harmonic flow vector is

$$Q_n \equiv \sum_{j=1}^{M} \exp(in\phi_j) \text{ with } i = \sqrt{-1}, \tag{3.1}$$

where ϕ_j is the azimuthal angle of the jth particle. The absolute square of the Q_n is

$$Q_n Q_n^* = \sum_{j,k=1}^{M} \exp in(\phi_j - \phi_k) = M + \sum_{j,k=1,j\neq k}^{M} \exp in(\phi_j - \phi_k). \tag{3.2}$$

The second term on the right-hand side of Eq. (3.2) is exactly the square of Eq. (1.5), that is, v_n^2. $\sum_{j,k=1,j\neq k} \exp in(\phi_j - \phi_k)$ divided by the number of pairs is called the second order moment,

$$\langle 2_n \rangle = \frac{Q_n Q_n^* - M}{M(M-1)}. \tag{3.3}$$

The single bracket represents the average over particles in one event. The second cumulant is equal to the average of the second moment over events.

$$V_n\{2\} = \langle\langle 2_n \rangle\rangle. \tag{3.4}$$

The double brackets represent the average over particles and events. The cumulant gives the estimation of flow v_n by

$$v_n\{2\} = \sqrt{V_n\{2\}}. \tag{3.5}$$

For higher order cumulants:

$$v_n\{4\} = \sqrt[4]{V_n\{4\}}; \tag{3.6}$$

$$v_n\{4\} = \sqrt[6]{\frac{1}{4} V_n\{6\}}, \tag{3.7}$$

where

$$V_n\{4\} = 2\langle\langle 2_n \rangle\rangle^2 - \langle\langle 4_n \rangle\rangle; \tag{3.8}$$

$$V_n\{6\} = \langle\langle 6_n \rangle\rangle - 9\langle\langle 2_n \rangle\rangle\langle\langle 4_n \rangle\rangle + 12\langle\langle 2_n \rangle\rangle^3. \tag{3.9}$$

$$\langle 4 \rangle_n = \frac{|Q_n|^4 + |Q_{2n}|^2 - 2 \cdot \text{Re}[Q_{2n} Q_n^* Q_n^*]}{M(M-1)(M-2)(M-3)}$$

$$-2\frac{2(M-2)\cdot|Q_n|^2 - M(M-3)}{M(M-1)(M-2)(M-3)};\tag{3.10}$$

$$\langle 6\rangle_n = \frac{|Q_n|^6 + 9\cdot|Q_{2n}|^2|Q_n|^2 - 6\cdot \text{Re}[Q_{2n}Q_nQ_n^*Q_n^*Q_n^*]}{M(M-1)(M-2)(M-3)(M-4)(M-5)}$$

$$+4\frac{\text{Re}[Q_{3n}Q_n^*Q_n^*Q_n^*] - 3\cdot \text{Re}[Q_{3n}Q_{2n}^*Q_n^*]}{M(M-1)(M-2)(M-3)(M-4)(M-5)}$$

$$+2\frac{9(M-4)\cdot \text{Re}[Q_{2n}Q_n^*Q_n^*] + 2\cdot|Q_{3n}|^2}{M(M-1)(M-2)(M-3)(M-4)(M-5)}$$

$$-9\frac{|Q_n|^4 + |Q_{2n}|^2}{M(M-1)(M-2)(M-3)(M-5)}$$

$$+18\frac{|Q_n|^2}{M(M-1)(M-3)(M-4)}$$

$$-\frac{6}{(M-1)(M-2)(M-3)}.\tag{3.11}$$

As the above equations show, the nested loops are now replaced by means of single particle quantities, which need only a single loop over particles to calculate. The complex calculation is largely simplified and the computing time is reduced.

Flow is a global property of particle anisotropic distributions in the whole event, with respect to a common harmonic plane (the participant plane). Nonflow is intrinsic particle correlations unrelated to the common plane, such as jet correlations, resonance particle decays, and quantum statistics [5–7]. These nonflow correlations contaminate flow measurements.

Nonflow is mostly a short range correlation in $\Delta\phi$ and $\Delta\eta$. In order to eliminate the contribution from nonflow, a $\Delta\eta$-gap method is usually used for the two-particle cumulant measurement. A large $\Delta\eta$ separation eliminates short range correlations, hence, minimizes the nonflow contribution [8]. The STAR TPC (with acceptance $-1 < \eta < 1$) can be used as an example to illustrate the $\Delta\eta$-gap two-particle cumulant flow measurement method. The $\Delta\eta$-gap can be chosen as $|\Delta\eta| > 1$. Two symmetrical η windows with respect to zero rapidity are chosen for the two-particle cumulant measurement such that one particle lies in $-1 < \eta < -0.5$, named group A, and the other particle in $0.5 < \eta < 1$, named group B. The $\Delta\eta$-gap size is determined by two factors. One is the typical angular size of the nonflow, such as the size of the near-side jet cone [9]. The other factor is good statistics needed for the cumulant results. In symmetric Au + Au collisions, the physics quantities in the two groups A and B should be the identical, when the group A and B are symmetric about zero rapidity. The second order harmonic moment for one particle in A and the other particle in B for integrated p_T (particles in all p_T) is

$$\langle\langle 2_n\rangle\rangle(A;B) = \frac{Q_n(A)Q_n^*(B)}{M(A)M(B)}.\tag{3.12}$$

The two-particle flow estimate in window η_A and/or η_B is

$$v_n(\eta_{A,B}) = \sqrt{\langle\langle 2_n \rangle\rangle (A; B)}. \tag{3.13}$$

Now suppose that group C is the particle of interest (e.g., within a particular p_T bin to study the p_T-dependence of v_n) and group A is the reference particle. A $\Delta\eta$-gap is often also applied between group C and group A. The v_n of the particle of interest (group C) can be calculated with the following equation:

$$v_n(C) = \frac{\langle\langle 2_n \rangle\rangle (C; A)}{\sqrt{\langle\langle 2_n \rangle\rangle (A; B)}}. \tag{3.14}$$

3.2 Data Sample and Analysis Cuts

A total of 19 million minimum bias Au + Au events at $\sqrt{s_{NN}} = 200\,\mathrm{GeV}$ has been collected by STAR and are used in this thesis. The main detector, the TPC, covers the pseudo-rapidity $-1 < \eta < 1$. The minimum bias collisions are triggered by the ZDC and the Beam-Beam Counter (BBC) detectors [3]. The centrality is measured by the charged particle multiplicity in $-0.5 < \eta < 0.5$ in the TPC. The number of participant nucleons N_{part} and the number of binary nucleon-nucleon collisions N_{coll} for the corresponding Au + Au collisions at $\sqrt{s_{NN}} = 200\,\mathrm{GeV}$ are listed in [10]. The collision point is required to be within 30 cm of the TPC center along the beam pipe to assure a uniform acceptance. The charged particle tracks are reconstructed by fitting the hit points in the TPC. The number of hit points is required to be larger than 51 % of the number of possible hit points, and also more than 20 hits out of the maximum 45 detecting points. The 51 % requirement is to avoid split tracks (single track is mistakenly reconstructed as two tracks close together), and the 20 points requirement is to assure there are enough points for a good fit. The dca to the primary collision vertex is required to be less than 2 cm.

3.3 v_3 Measurement Result

Figure 3.1 shows the elliptic flow and third harmonic flow from two- and four-particle Q-cumulant [2] measurements as a function of centrality for Au + Au collisions at 200 GeV. A $\Delta\eta$-gap $|\Delta\eta| > 1$ is applied. The elliptic flow, v_2, reaches its minimum in the most central collisions. In these collisions, the two Au nuclei overlap on top of each other such that the collision region is approximately a round shape. The initial eccentricity ϵ_2 of the collision system is small, which leads to a small elliptic flow for the final particles. As the centrality decreases, the eccentricity and the v_2 increase. The v_2 reaches its maximum in collisions of 40–50 % centrality class. As for more peripheral collisions, the overlap area decreases, and the energy density becomes small, so less flow is developed.

Fig. 3.1 Two-particle Q-cumulant flow for the second harmonic $v_2\{2\}$ (*black dots*) and the triangular $v_3\{2\}$ (*green stars*), and four-particle Q-cumulant flow for the second harmonic $v_2\{4\}$ (*red squares*) as a function of centrality in Au + Au at 200 GeV

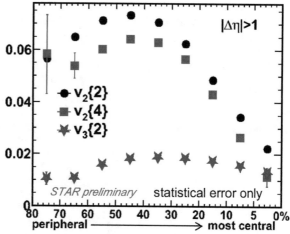

Fig. 3.2 Two-particle Q-cumulant flow for the third harmonic $v_3\{2\}$ with $|\Delta\eta|$-gap $|\Delta\eta| > 1$ (*pink stars*), $|\Delta\eta| > 0.7$ (*blue stars*) and $|\Delta\eta| > 0$ (*green stars*) as a function of centrality in Au + Au at 200 GeV

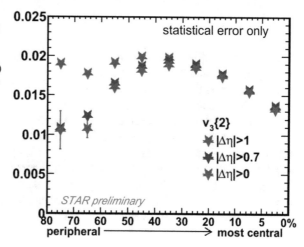

The four-particle flow is smaller than the two-particle flow, because the nonflow contribution, which is a few-body correlation, is suppressed in the four-particle flow. In addition, the flow fluctuation contribution is negative in the four-particle flow measurement, while it is positive in two-particle flow (see Chap. 4).

The magnitude of the third harmonic flow is less than the elliptic flow v_2. The v_3 dependence on centrality is weak. Since v_3 is due to event-by-event fluctuations, it is not expected to be primarily correlated with the reaction plane, making it unlikely that v_3 will follow the strong centrality dependence as v_2. Further discussions on the correlation between different harmonic event planes (the experimental estimated Ψ_n) can be found in [11].

Figure 3.2 compares the two-particle v_3 centrality dependence measurements with three $\Delta\eta$-gaps: $|\Delta\eta| > 1$ for one particle in $-1 < \eta < -0.5$ and the other in $0.5 < \eta < 1$; $|\Delta\eta| > 0.7$ for one in $-1 < \eta < -0.35$ and the other in

Fig. 3.3 Two-particle Q-cumulant flow for the third harmonic $v_3\{2\}$ with $\Delta\eta$-gap $|\Delta\eta| > 0.7$ in 0–5 % (*pink dots*) and 30–40 % (*blue dots*) as a function of transverse momentum p_T in Au + Au at 200 GeV. The *dashed curves* are the event-by-event hydrodynamic calculation [12]

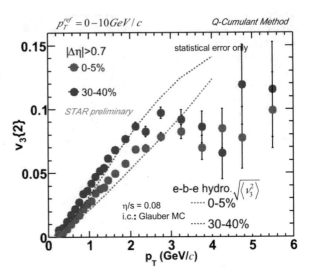

$0.35 < \eta < 1; |\Delta\eta| > 0$ for one in $-1 < \eta < 0$ and the other in $0 < \eta < 1$. $\Delta\eta$-gap size of 1 and 0.7 give similar v_3 for all centralities. In peripheral collisions, $|\Delta\eta| > 0$ is larger than those two with larger $\Delta\eta$-gap. Their difference decreases as the centrality decreases.

Figure 3.3 shows the measured v_3 as a function of transverse momentum p_T for two centrality classes, 0–5 % and 30–40 %. For both centrality classes, v_3 increases with p_T and saturates above 2–3 GeV/c. The event-by-event hydrodynamic calculations [12] are superimposed as the dashed curves. The theoretic calculation chooses the viscosity to entropy density ratio to be $\eta/s = 0.08$ with the Glauber initial state condition. The hydrodynamic calculation gives a good description for the v_3 trend below 2 GeV/c. Above 2 GeV/c, hydrodynamics may no longer be considered applicable. However, the flow measurement may have nonflow contamination which is not present in theoretical calculation. Therefore nonflow contamination needs to be isolated from flow measurement for a precise comparison (see Chap. 4).

References

1. B. Alver, G. Roland, Collision-geometry fluctuations and triangular flow in heavy-ion collisions. Phys. Rev. C. **81**, 054905 (2010)
2. A. Bilandzic, R. Snellings, S. Voloshin, Flow analysis with cumulants: direct calculations. Phys. Rev. C **83**, 044913 (2011)
3. K. Ackermann et al., STAR detector overview. Nucl. Instrum. Methods **A499**(2–3), 624–632 (2003)
4. L. Adamczky et al., Third harmonic flow of charged particles in Au+Au collisions at $\sqrt{s_{NN}} =$ 200 GeV. Phys. Rev. C **88**, 014904 (2013)

5. N. Borghini, P.M. Dinh, J.Y. Ollitrault, Is the analysis of flow at the CERN super proton synchrotron reliable?. Phys. Rev. C **62**, 034902 (2000)
6. N. Borghini, P.M. Dinh, J. Y. Ollitrault, New method for measuring azimuthal distributions in nucleus-nucleus collisions. Phys. Rev. C **63**, 054906 (2001)
7. N. Borghini, P.M. Dinh, J.Y. Ollitrault, Flow analysis from multiparticle azimuthal correlations. Phys. Rev. C **64**, 054901 (2001)
8. H. Agakishiev et al., Event-plane-dependent dihadron correlations with harmonic v_n subtraction in Au + Au collisions at $\sqrt{s_{NN}} = 200$ GeV. Phys. Rev. C **89**, 041901 (2014)
9. M. Procura, W.J. Waalewijn, Fragmentation in jets: cone and threshold effects. Phys. Rev. D **85**, 114041 (2012)
10. G. Agakishiev et al., Energy and system-size dependence of two- and four-particle v_2 measurements in heavy-ion collisions at $\sqrt{s_{NN}} = 62.4$ and 200 GeV and their implications on flow fluctuations and nonflow. Phys. Rev. C **86**, 014904 (2012)
11. G. Aad et al., Measurement of event-plane correlations in $\sqrt{s_{NN}} = 2.76$ TeV lead-lead collisions with the ATLAS detector. Phys. Rev. C **90**, 024905 (2014)
12. B. Schenke, S. Jeon, C. Cale, Elliptic and triangular flow in event-by-event $D = 3 + 1$ viscous hydrodynamics. Phys. Rev. Lett. **106**, 042301 (2011)

Chapter 4
Isolation of Flow and Nonflow Correlations

This chapter of thesis work has been summarized in [7].

Through the comparison of anisotropic flow measurements and hydrodynamic calculations, properties of the early stage of the collision system may be extracted. One of the important conclusions from the comparison is that the ratio of the shear viscosity to entropy density of the QGP, η/s, was found to be not much larger than the conjectured quantum limit of $1/4\pi$ [8]. This is where the name of "perfect fluid" comes from for QGP.

The momentum-space anisotropic flow can be characterized by the Fourier coefficients v_n as shown in Eq. (1.3). The participant plane Ψ_n is unaccessible experimentally. The anisotropic flow is thus measured by particle correlations, such as two-particle or multiple-particle correlations [2–4, 9]. (For event plane method, the event plane is first determined from all particles. Hence, it is effectively a two-particle correlation method.) Therefore, the flow measurement is contaminated by intrinsic particle correlations unrelated to the participant plane. Those correlations are generally called nonflow and are mainly due to jet fragmentation, final state Coulomb and strong interactions, resonance decays [2–4].

The two-particle correlation is given by Eq. (1.4). With non-zero nonflow, $V_n\{2\} = v_{n,\alpha} \cdot v_{n,\beta} + \delta_n$, where $v_{n,\alpha}, v_{n,\beta}$ stand for the flow of two particles at α and β, and δ_n is the nonflow contribution. Since the shear viscosity extracted from the flow measurement is sensitive to even a small change in flow value [10], it is important to eliminate nonflow contributions in flow measurements.

This chapter describes a method to separate flow and nonflow in a data-driven way, with minimal reliance on models. Two- and four-particle cumulants with different pseudo-rapidity (η) combinations are measured. By exploiting the symmetry of the average flow $\langle v \rangle$ in η about zero rapidity in symmetric Au + Au collisions, $\Delta\eta$-independent and $\Delta\eta$-dependent contributions are separated. Flow, as an event-wise multiple-body azimuthal correlation, reflecting properties on the single-particle level, depends on η. However, nonflow is a few-body azimuthal

© Springer Science+Business Media New York 2016
L. Yi, *Study of Quark Gluon Plasma By Particle Correlations in Heavy Ion Collisions*, Springer Theses, DOI 10.1007/978-1-4939-6487-1_4

correlation which depends on the $\Delta\eta$ distance between the particles with intrinsic correlations. Therefore, the $\Delta\eta$-independent part can be associated with flow, while the $\Delta\eta$-dependent part can be associated with nonflow.

4.1 Analysis Method

In this analysis, the anisotropy was calculated using the two- and four-particle Q-cumulant method with unity weight for event average [1]. The non-uniform acceptance corrected is applied (see Sect. 4.2.2). The Q-cumulant method allows the calculation of multi-particle cumulant with only one loop over particles without the nested loop over pair or high multiplet (Sect. 3.1).

The two-particle cumulant, with one particle at pseudo-rapidity η_α and another at η_β, illustrated in Fig. 4.1, is [11, 12]

$$V\{2\} \equiv \langle\langle e^{i(\phi_\alpha - \phi_\beta)} \rangle\rangle = \langle v(\eta_\alpha)v(\eta_\beta)\rangle + \delta(\Delta\eta)$$
$$\equiv \langle v(\eta_\alpha)\rangle\langle v(\eta_\beta)\rangle + \sigma(\eta_\alpha)\sigma(\eta_\beta) + \sigma'(\Delta\eta) + \delta(\Delta\eta), \qquad (4.1)$$

where $\Delta\eta = |\eta_\beta - \eta_\alpha|$. The single brackets represent the average over events only, while the double brackets are for the average over particle pairs and the average over events. The harmonic number n is suppressed to lighten the notation. The average flow $\langle v \rangle$ is the anisotropy parameter with respect to the participant plane. σ is the flow fluctuation of $\langle v \rangle$. Since average flow reflects the property on the single-particle level, $\langle v \rangle$ and σ are only functions of η, which are $\Delta\eta$-independent. However, a $\Delta\eta$-dependent flow fluctuation component could exist. The harmonic planes Ψ_n could depend on η due to the initial energy density longitudinal fluctuation [13–15]. The σ' is used to denote such $\Delta\eta$-dependent part of the flow fluctuation. The δ is nonflow, which is generally a function of $\Delta\eta$, but may also depend on η.

For the four-particle cumulant, two particles are chosen to be at η_α and another two are at η_β. For easier discussion, the square root of the four-particle cumulant is used in calculation, which has the same order in $\langle v \rangle$ as the two-particle cumulant. The square root of the four-particle cumulant is given by

$$V^{\frac{1}{2}}\{4\} \equiv \sqrt{\langle\langle e^{i(\phi_\alpha + \phi_\alpha - \phi_\beta - \phi_\beta)} \rangle\rangle}$$
$$\approx \langle v(\eta_\alpha)\rangle\langle v(\eta_\beta)\rangle - \sigma(\eta_\alpha)\sigma(\eta_\beta) - \sigma'(\Delta\eta), \qquad (4.2)$$

Fig. 4.1 Illustration of one pair of two-particle cumulant $V(\eta_\alpha, \eta_\beta)$ with one particle at η_α and the other particle at η_β

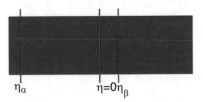

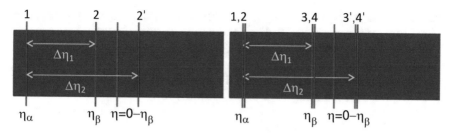

Fig. 4.2 *Left*: Two pair of two-particle cumulants, $V\{2\}(\eta_\alpha, \eta_\beta)$ with one particle at η_α and the other particle at η_β, and $V\{2\}(\eta_\alpha, -\eta_\beta)$ with one particle at η_α and the other particle at $-\eta_\beta$. One pair is denoted as particle 1 and 2. The other pair is particle 1 and 2'. *Right*: Two pair of four-particle cumulants, $V\{4\}(\eta_\alpha, \eta_\beta)$ with two particles at η_α and the other two particles at η_β, and $V\{4\}(\eta_\alpha, \eta_\beta)$ with two particles at η_α and the other two particles at $-\eta_\beta$. One quadruplet is denoted as particles 1, 2, 3, and 4. The other quadruplet is 1, 2, 3', and 4'

where the assumption for the approximation is that the flow fluctuation is relatively small compared with the average flow [16]. The flow fluctuation has negative contribution in $V^{1/2}\{4\}$, while it is positive in $V\{2\}$. The contribution of two-particle nonflow effect to $V^{1/2}\{4\}$ is suppressed. The contribution of four-particle nonflow effect to $V^{1/2}\{4\}$ is $\propto 1/M^3$ (M is multiplicity) and is, therefore, negligible [17, 18].

The analysis method described in Ref. [12] is used to extract the values of the average flow, $\langle v \rangle$, the $\Delta\eta$-dependent and $\Delta\eta$-independent flow fluctuations, σ' and σ, and the nonflow contribution, δ. Consider two pairs of two-particle cumulants as Fig. 4.2 left panel shows. The difference between their cumulants $V\{2\}$ at $(\eta_\alpha, \eta_\beta)$ and $(\eta_\alpha, -\eta_\beta)$ is

$$
\begin{aligned}
\Delta V\{2\} &\equiv V\{2\}(\eta_\alpha, \eta_\beta) - V\{2\}(\eta_\alpha, -\eta_\beta) \\
&\equiv V\{2\}(\Delta\eta_1) - V\{2\}(\Delta\eta_2) \\
&= \Delta\sigma' + \Delta\delta,
\end{aligned}
\tag{4.3}
$$

where $\eta_\alpha < \eta_\beta < 0$ or $0 < \eta_\beta < \eta_\alpha$ is required. Similarly, for two pairs of four-particle cumulants as illustrated in Fig. 4.2 right panel, the difference for $V^{1/2}\{4\}$ is,

$$
\begin{aligned}
\Delta V^{\frac{1}{2}}\{4\} &\equiv V^{\frac{1}{2}}\{4\}(\eta_\alpha, \eta_\beta) - V^{\frac{1}{2}}\{4\}(\eta_\alpha, -\eta_\beta) \\
&\equiv V^{\frac{1}{2}}\{4\}(\Delta\eta_1) - V^{\frac{1}{2}}\{4\}(\Delta\eta_2) \\
&\approx -\Delta\sigma'.
\end{aligned}
\tag{4.4}
$$

Here $\Delta\eta_1 \equiv \eta_\beta - \eta_\alpha$, $\Delta\eta_2 \equiv -\eta_\beta - \eta_\alpha$, $\Delta\sigma' = \sigma'(\Delta\eta_1) - \sigma'(\Delta\eta_2)$, and $\Delta\delta = \delta(\Delta\eta_1) - \delta(\Delta\eta_2)$. In symmetric Au + Au collisions, the two $\Delta\eta$-independent terms in Eqs. (4.1) and (4.2) have $\langle v(\eta_\alpha) \rangle \langle v(\eta_\beta) \rangle = \langle v(\eta_\alpha) \rangle \langle v(-\eta_\beta) \rangle$

and $\sigma(\eta_\alpha)\sigma(\eta_\beta) = \sigma(\eta_\alpha)\sigma(-\eta_\beta)$. The differences from these $\Delta\eta$-independent terms have zero contribution in Eqs. (4.3) and (4.4). Therefore the differences in Eqs. (4.3) and (4.4) depend only on the $\Delta\eta$-dependent terms: flow fluctuation $\Delta\sigma'$ and nonflow $\Delta\delta$.

The goal is to parameterize the flow fluctuation $\Delta\sigma'$ and nonflow $\Delta\delta$. The plan is as follows: extract the empirical functional form for

$$D(\Delta\eta) = \sigma'(\Delta\eta) + \delta(\Delta\eta), \qquad (4.5)$$

from $\Delta V_2\{2\}$ data; obtain the σ' result from $\Delta V_2^{1/2}\{4\}$; use D and σ' to determine δ; finally $\langle v \rangle$ and σ.

4.2   Data Analysis

4.2.1   Data Samples and Analysis Cuts

This thesis work primarily uses data taken by the STAR TPC [19]. A total of 25 million Au + Au collisions at $\sqrt{s_{NN}} = 200\,\text{GeV}$, collected with a minimum bias trigger in 2004, are used. The collisions are selected to have a primary event vertex within $|z_{vtx}| < 30\,\text{cm}$ along the beam axis to ensure nearly uniform detector acceptance. The centrality definition is based on the raw charged particle multiplicity within $|\eta| < 0.5$ in TPC [6]. The charged particle tracks used in this analysis are required to satisfy the following conditions: the transverse momentum $0.15 < p_T < 2\,\text{GeV}/c$ to minimize jet contributions; the distance of closest approach to the event vertex $|dca| < 3\,\text{cm}$ to select the particles from the primary collision vertex and exclude the ones from a secondary particle decay vertex; the number of fit points along the track greater than 20, and the ratio of the number of fit points along the track to the maximum number of possible fit points larger than 51 % for good primary track reconstruction and to avoid split tracks [20]. For the particles used in this thesis work, the particle pseudo-rapidity is restricted to $|\eta| < 1$ for a high tracking efficiency.

4.2.2   Non-uniform Acceptance Correction

For a detector with uniform acceptance, the terms such as $\langle\langle\cos\phi_\alpha\rangle\rangle$ and $\langle\langle\sin\phi_\alpha\rangle\rangle$ vanish. Although STAR TPC has a nearly uniform acceptance, there could still be some deficiencies. The non-uniform acceptance can be corrected in the following way [1]:

$$V\{2\} = \langle\langle 2\rangle\rangle - [\langle\langle\cos\phi_\alpha\rangle\rangle \cdot \langle\langle\cos n\phi_\beta\rangle\rangle + \langle\langle\sin\phi_\alpha\rangle\rangle \cdot \langle\langle\sin\phi_\beta\rangle\rangle]; \qquad (4.6)$$

$$V\{4\} = \langle\langle 4 \rangle\rangle - 2\langle\langle 2 \rangle\rangle^2 - 6\langle\langle \cos\phi_\alpha \rangle\rangle^3 \langle\langle \cos\phi_\beta \rangle\rangle + 2\langle\langle \cos\phi_\alpha \rangle\rangle^2 \langle\langle \cos(\phi_\beta + \phi_\alpha) \rangle\rangle$$
$$- 2\langle\langle \cos\phi_\alpha \rangle\rangle\langle\langle \cos(\phi_\beta + \phi_\alpha - \phi_\alpha) \rangle\rangle + 4\langle\langle \cos\phi_\alpha \rangle\rangle^2 \langle\langle \cos(\phi_\beta - \phi_\alpha) \rangle\rangle$$
$$- \langle\langle \cos\phi_\alpha \rangle\rangle\langle\langle \cos(\phi_\beta - \phi_\alpha - \phi_\alpha) \rangle\rangle + 4\langle\langle \cos\phi_\alpha \rangle\rangle\langle\langle \cos\phi_\beta \rangle\rangle$$
$$\times \langle\langle \cos(\phi_\alpha - \phi_\alpha) \rangle\rangle - \langle\langle \cos\phi_\beta \rangle\rangle\langle\langle \cos(\phi_\alpha - \phi_\alpha - \phi_\alpha) \rangle\rangle$$
$$+ 2\langle\langle \cos\phi_\alpha \rangle\rangle\langle\langle \cos\phi_\beta \rangle\rangle\langle\langle \cos n(-\phi_\alpha - \phi_\alpha) \rangle\rangle - \langle\langle \cos(\phi_\beta + \phi_\alpha) \rangle\rangle$$
$$\times \langle\langle \cos(-\phi_\alpha - \phi_\alpha) \rangle\rangle - 6\langle\langle \cos\phi_\alpha \rangle\rangle\langle\langle \cos\phi_\beta \rangle\rangle\langle\langle \sin\phi_\alpha \rangle\rangle^2$$
$$- 2\langle\langle \cos(\phi_\beta + \phi_\alpha) \rangle\rangle\langle\langle \sin\phi_\alpha \rangle\rangle^2 + 4\langle\langle \cos(\phi_\beta - \phi_\alpha) \rangle\rangle\langle\langle \sin\phi_\alpha \rangle\rangle^2$$
$$- 6\langle\langle \cos\phi_\alpha \rangle\rangle^2 \langle\langle \sin\phi_\alpha \rangle\rangle\langle\langle \sin\phi_\beta \rangle\rangle + 4\langle\langle \cos(\phi_\alpha - \phi_\alpha) \rangle\rangle\langle\langle \sin\phi_\alpha \rangle\rangle\langle\langle \sin\phi_\beta \rangle\rangle$$
$$- 2\langle\langle \cos n(-\phi_\alpha - \phi_\alpha) \rangle\rangle\langle\langle \sin\phi_\alpha \rangle\rangle\langle\langle \sin\phi_\beta \rangle\rangle$$
$$- 6\langle\langle \sin\phi_\alpha \rangle\rangle^3 \langle\langle \sin\phi_\beta \rangle\rangle + 4\langle\langle \cos\phi_\alpha \rangle\rangle\langle\langle \sin\phi_\alpha \rangle\rangle\langle\langle \sin(\phi_\beta + \phi_\alpha) \rangle\rangle$$
$$- 2\langle\langle \sin\phi_\alpha \rangle\rangle\langle\langle \sin(\phi_\beta + \phi_\alpha - \phi_\alpha) \rangle\rangle + \langle\langle \sin\phi_\alpha \rangle\rangle\langle\langle \cos(\phi_\beta - \phi_\alpha - \phi_\alpha) \rangle\rangle$$
$$+ \langle\langle \sin\phi_\beta \rangle\rangle\langle\langle \sin(\phi_\alpha - \phi_\alpha - \phi_\alpha) \rangle\rangle - 2\langle\langle \cos\phi_\beta \rangle\rangle\langle\langle \sin\phi_\alpha \rangle\rangle$$
$$\times \langle\langle \sin(-\phi_\alpha - \phi_\alpha) \rangle\rangle - 2\langle\langle \cos\phi_\alpha \rangle\rangle\langle\langle \sin\phi_\beta \rangle\rangle\langle\langle \sin(-\phi_\alpha - \phi_\alpha) \rangle\rangle$$
$$+ \langle\langle \sin(\phi_\beta + \phi_\alpha) \rangle\rangle\langle\langle \sin(-\phi_\alpha - \phi_\alpha) \rangle\rangle. \tag{4.7}$$

The non-uniform acceptance correction dependence on centrality classes is summarized in Table 4.1. The correction for 20–30 % centrality is 0.7 % for the second harmonic two-particle cumulant $V_2\{2\}$, and 0.5 % for the square root of the second harmonic four-particle cumulant $V_2^{1/2}\{4\}$. The largest acceptance correction is 1.8 % for $V_2\{2\}$ in the most central collisions, and 2 % for $V_2^{1/2}\{4\}$ in the most peripheral collisions.

4.2.3 Track Merging Effect

When two particle trajectories in the TPC are too close to each other, the tracking algorithm may mistakenly reconstruct their hit points together as one track. The deficiency of reconstruction of two close tracks is called the track merging effect.

Table 4.1 Non-uniform acceptance correction centrality dependence

	0–20 %	20–30 %	30–40 %	40–50 %	50–80 %
Relative $V_2\{2\}$ non-uniform correction	1.8 %	0.7 %	0.5 %	0.4 %	0.4 %
Relative $V_2\{4\}$ non-uniform correction	1.6 %	0.5 %	0.4 %	1 %	2 %

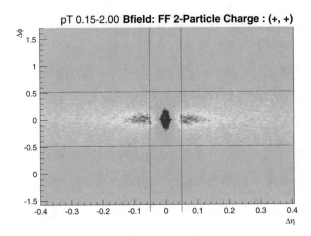

Fig. 4.3 The $\Delta\eta - \Delta\phi$ two-particle correlation demonstrating deficit due to track merging effect in Au + Au collisions at $\sqrt{s_{NN}} = 200$ GeV for particles in $0.15 < p_T < 2$ GeV/c. The x axis is the pseudo-rapidity separation $\Delta\eta$ between two particles and the y axis is their azimuthal angle difference $\Delta\phi$. The figure shows the correlation for two positive particles in the positive full magnetic field polarity (positive z direction). The *red* color represents higher number of pairs. The *blue* color represents lower number of pairs. The four red lines indicate $\Delta\eta$ and $\Delta\phi$ region affected by track merging deficit

The azimuthal opening angle $\Delta\phi$ for the particle pair affected by track merging effect depends on the particles' charges, transverse momenta, and the magnetic field polarity, while their pseudo-rapidity separation $\Delta\eta$ is independent of charges and magnetic field. The detailed study on track merging effect can be found at [21].

As Fig. 4.3 shows, for particles with $0.15 < p_T < 2$ GeV/c in Au + Au collisions at $\sqrt{s_{NN}} = 200$ GeV, the track merging effect is limited to two particles with $|\Delta\eta| < 0.05$. Since the track merging effect is localized in the region $|\Delta\eta| < 0.05$, the $V_n\{2\}$ and $V_n\{4\}$ points in the those region are excluded from further analysis.

4.2.4   Two- and Four-Particle Cumulant Measurements

The two- and four-particle cumulants were measured for various $(\eta_\alpha, \eta_\beta)$ pairs and quadruplets. Figure 4.4 shows the results for 20–30 % central Au + Au collisions. Panels (a) and (b) are the two-particle second and third harmonic cumulants, $V_2\{2\}(\eta_\alpha, \eta_\beta)$ and $V_3\{2\}(\eta_\alpha, \eta_\beta)$, respectively. Panel (c) is the square root of the four-particle second harmonic cumulant, $V_2^{1/2}\{4\}(\eta_\alpha, \eta_\alpha, \eta_\beta, \eta_\beta)$. Moving off the diagonal, the pseudo-rapidity separation between particle pairs $|\Delta\eta| = |\eta_\alpha - \eta_\beta|$ increases. $V_2\{2\}$ decreases as $|\Delta\eta|$ increases as shown in Fig. 4.4. Similar trend is observed for $V_3\{2\}$, while $V_3\{2\}$ magnitude is smaller, and decreases more rapidly with $\Delta\eta$ than $V_2\{2\}$ does. The four-particle cumulant $V_2^{1/2}\{4\}$ is roughly

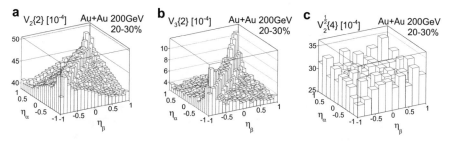

Fig. 4.4 The second (**a**) and third (**b**) harmonic two-particle cumulants for $(\eta_\alpha, \eta_\beta)$ pairs and the second harmonic four-particle cumulant (**c**) for $(\eta_\alpha, \eta_\alpha, \eta_\beta, \eta_\beta)$ quadruplets for 20–30 % central Au + Au collisions at $\sqrt{s_{NN}} = 200$ GeV. Figure reprinted from [7] under Creative Commons Attribution 4.0 License http://creativecommons.org/licenses/by/4.0/

constant and the magnitude is smaller than two-particle cumulant $V_2\{2\}$ as expected since $V_2^{1/2}\{4\}$ is less affected by the nonflow and the flow fluctuation has negative contribution in $V_2^{1/2}\{4\}$. The points affected by track merging on the diagonal are (see Sect. 4.2.3) excluded from analysis.

4.2.5 Nonflow Parameterization

Figure 4.5a shows the measured two-particle second harmonic cumulant difference $\Delta V_2\{2\}$. For each $\Delta\eta_1$ (the points with same color), the data value appears to be linear in $\Delta\eta_2 - \Delta\eta_1$ except near $\Delta\eta_1 = \Delta\eta_2$, as shown by dashed lines in Fig. 4.5a, b. Moreover, the magnitude of $\Delta V_2\{2\}$ decreases with increasing $\Delta\eta_1$. Linear fits represented by the dashed lines indicate that the intercept decreases exponentially with increasing $\Delta\eta_1$, but the slopes are all similar. Such behavior of $\Delta V_2\{2\}$ could be described mathematically as $a\exp(-\frac{\Delta\eta_1}{b}) + k(\Delta\eta_2 - \Delta\eta_1)$. The mathematical function form of $V_2\{2\}$ needs to be obtained from $\Delta V_2\{2\}$. In order to express the measured $\Delta V_2\{2\}$ in the form of $D(\Delta\eta_1) - D(\Delta\eta_2) = (\sigma'(\Delta\eta_1) + \delta(\Delta\eta_1)) - (\sigma'(\Delta\eta_2) + \delta(\Delta\eta_2)) = (\sigma'(\Delta\eta_1) - \sigma'(\Delta\eta_2)) + (\delta(\Delta\eta_1) - \delta(\Delta\eta_2))$, two improvements are made to the initial guess of the exponential + linear form of function $D(\Delta\eta)$. Firstly, a term $a\exp(-\frac{\Delta\eta_2}{b})$ is added which is small when $\Delta\eta_2$ is significantly larger than $\Delta\eta_1$. Secondly, because the linear term is unbounded, it is replaced by the subtraction of two wide Gaussian terms. The Gaussian functions approach zero when $\Delta\eta_1$ or $\Delta\eta_2$ becomes large, consistent with the behavior of nonflow. The measured two-particle cumulant difference can then be described by Eq. (4.9):

$$D(\Delta\eta) = a\exp\left(-\frac{\Delta\eta}{b}\right) + A\exp\left(-\frac{\Delta\eta^2}{2\sigma^2}\right), \qquad (4.8)$$

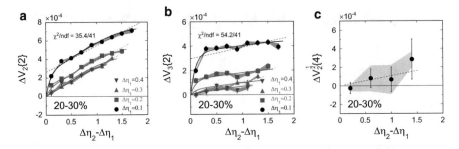

Fig. 4.5 The (**a**) $V_2\{2\}$ and (**b**) $V_3\{2\}$ difference between the pairs at $(\eta_\alpha, \eta_\beta)$ and $(\eta_\alpha, -\eta_\beta)$. The *dashed lines* are linear fits for each data set of $\Delta\eta_1$ value separately. The *solid curves* are a single fit of Eq. (4.8) to all data points with different $\Delta\eta_1$. (**c**) The $V_2^{1/2}\{4\}$ difference between quadruplets at $(\eta_\alpha, \eta_\alpha, \eta_\beta, \eta_\beta)$ and $(\eta_\alpha, \eta_\alpha, -\eta_\beta, -\eta_\beta)$. The *dashed line* is a linear fit to the data points. The *gray band* represents the systematic error. The data are from 20–30 % central Au + Au collisions at $\sqrt{s_{NN}} = 200$ GeV. Figure reprinted from [7] under Creative Commons Attribution 4.0 License http://creativecommons.org/licenses/by/4.0/

such that

$$\Delta V\{2\} = D(\Delta\eta_1) - D(\Delta\eta_2)$$

$$= \left[a\exp\left(\frac{-\Delta\eta_1}{b}\right) + A\exp\left(\frac{-\Delta\eta_1^2}{2\sigma^2}\right) \right]$$

$$- \left[a\exp\left(\frac{-\Delta\eta_2}{b}\right) + A\exp\left(\frac{-\Delta\eta_2^2}{2\sigma^2}\right) \right], \qquad (4.9)$$

follows from Eq. (4.3). Function $D(\Delta\eta)$ has four parameters $a, A, b,$ and σ. These parameters were determined by simultaneously fitting all $V_2\{2\}$ data points with function (4.9). The fit results are shown in Fig. 4.5a as the solid curves with $\chi^2/\text{ndf} \approx 1$. The same procedure was repeated for the third harmonic $V_3\{2\}$ as shown in Fig. 4.5b. The D function describes the $\Delta\eta$-dependent part of the two-particle cumulant. The procedure of obtaining the form of D function is data-driven.

A similar procedure is performed on the measured difference of the square root of the four-particle cumulant, Eq. (4.2). $\Delta V_2^{1/2}\{4\} = \sigma'(\Delta\eta_1) - \sigma'(\Delta\eta_2)$ is fit by a linear function $k'(\Delta\eta_2 - \Delta\eta_1)$, as the dashed line shows in Fig. 4.5c. The slope k' from the fit is $(1.1 \pm 0.8) \times 10^{-4}$. In Fig. 4.5c, each data point is the average of $\Delta V_2^{1/2}\{4\}$ for all $\Delta\eta_1$ at same $\Delta\eta_2 - \Delta\eta_1$ value. With the $\sigma'(\Delta\eta)$ result from the $\Delta V_2^{1/2}\{4\}$ parameterization, the contribution from nonflow, δ, can then be determined through Eq. (4.5).

Subtracting the parameterized D of Eq. (4.8) from the measured two-particle cumulants, $V\{2\}$ of Eq. (4.1) yields the $\Delta\eta$-independent terms $\langle v^2 \rangle \equiv \langle v \rangle^2 + \sigma^2$. Together with $V^{1/2}\{4\}$ of Eq. (4.2), the values of $\langle v \rangle$ and σ may be individually determined.

4.3 Results and Discussion

Figure 4.6a, b shows the decomposed flow with $\Delta\eta$-independent flow fluctuations $\langle v(\eta_\alpha)v(\eta_\beta)\rangle$ [see Eq. (4.1)] for v_2 and v_3, respectively. The results are found to be constant over η in the measured pseudo-rapidity range $|\eta| < 1$. The observed decrease of $V\{2\}$ in Fig. 4.4 with increasing $\Delta\eta$ off diagonal is due to contributions from nonflow and $\Delta\eta$-dependent fluctuations. Note that there is no assumption about the η dependence of flow in the analysis; the flow can be $\Delta\eta$-independent but η-dependent. The observation that the decomposed flow and flow fluctuations are independent of η is, therefore, significant.

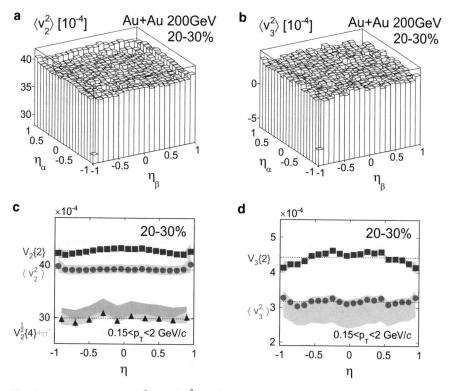

Fig. 4.6 The decomposed $\langle v^2\rangle = \langle v\rangle^2 + \sigma^2$ for the second (**a**) and third (**b**) harmonics for $(\eta_\alpha, \eta_\beta)$ pairs. (**c**): The two- and four-particle cumulants, $V_2\{2\}$ (*solid red squares*) and $V_2^{1/2}\{4\}$ (*solid blue triangles*), and the decomposed $\langle v_2^2\rangle$ (*solid green dots*) as a function of η for one particle while averaged over η of the partner particle. The *cyan band* on top of $V_2^{1/2}\{4\}$ points present $V_2^{1/2}\{4\} + \sigma'$. (**d**): $V_3\{2\}$ (*solid red squares*) and $\langle v_3^2\rangle$ (*solid green dots*) as a function of η. The *dashed lines* are the mean values averaged over η for 20–30 % central Au+Au collisions at $\sqrt{s_{NN}} = 200$ GeV. Figure reprinted from [7] under Creative Commons Attribution 4.0 License http://creativecommons.org/licenses/by/4.0/

Figure 4.6c, d shows the projections of $\langle v(\eta_\alpha)v(\eta_\beta)\rangle$ from Fig. 4.6a, b onto one η dimension. The shaded band shows the systematic uncertainty, dominated by the systematic errors in the subtracted $D(\Delta\eta)$ term. The projection of the $V_2\{2\}$ is shown as the red squares, where the shaded band is the systematic uncertainty. The projections are the respective quantities as a function of η of one particle averaged over all η of the other particle with unity weight. The flows with $\Delta\eta$-independent fluctuation averaged over η are $\sqrt{\langle v_2^2\rangle} = 6.27\% \pm 0.003\%(\text{stat.})^{+0.08}_{-0.07}\%(\text{sys.})$ and $\sqrt{\langle v_3^2\rangle} = 1.78\% \pm 0.008\%(\text{stat.})^{+0.09}_{-0.16}\%(\text{sys.})$ for the p_T range $0.15 < p_T < 2\,\text{GeV}/c$ in the 20–30% centrality Au + Au collisions. The quoted statistical errors are from the $V\{2\}$ measurements, while the systematic errors are dominated by the parameterization of D. The difference between $V\{2\}$ and $\langle v^2\rangle$ in Fig. 4.6c represents the $D(\eta)$ value versus η of one particle averaged over all η of the other particle. The projection of $V_2^{1/2}\{4\}$ is shown as the solid blue triangles. $V_2^{1/2}\{4\}$ is independent of η. The cyan band is $V_2^{1/2}\{4\} + \sigma' = \langle v\rangle^2 - \sigma^2$, with the systematic uncertainty dominated by the fitting uncertainty in σ'. The difference between the decomposed $\langle v^2\rangle = \langle v\rangle^2 + \sigma^2$ and $V_2^{1/2}\{4\} + \sigma' = \langle v\rangle^2 - \sigma^2$ is two times the $\Delta\eta$-independent flow fluctuation σ^2, which is also constant over η within the measured acceptance. The relative $\Delta\eta$-independent elliptic flow fluctuation is therefore

$$\frac{\sigma_2}{\langle v_2\rangle} = \sqrt{\frac{\langle v_2^2\rangle - (V_2^{\frac{1}{2}}\{4\} + \sigma')}{\langle v_2^2\rangle + (V_2^{\frac{1}{2}}\{4\} + \sigma')}}$$

$$= 34\% \pm 2\%(\text{stat.}) \pm 3\%(\text{sys.}),\qquad(4.10)$$

where the systematic error is dominated by uncertainty in the parameterization of σ'. The relative fluctuation result is consistent with the one from the PHOBOS experiment [22] and the previous STAR upper limit measurement [6].

A $\Delta\eta$-gap is usually applied to reduce nonflow contamination in flow measurements (Chap. 3). The nonflow $\bar{D}(|\Delta\eta|)$ contribution in the $\Delta\eta$-gap measurement is

$$\bar{D}(|\Delta\eta|) = \frac{\int_{|\Delta\eta|}^{2} d\Delta\eta' D(\Delta\eta')}{2 - |\Delta\eta|}.\qquad(4.11)$$

$|\Delta\eta| = 2$ is the acceptance limit of the TPC. $\bar{D}$ is the average of D with $|\Delta\eta|$ larger than a certain value. Figure 4.7a, b shows $\bar{D}(|\Delta\eta|)$ as a function of $\Delta\eta$-gap $|\Delta\eta| > x$ (x is the x-axis value) for the second and third harmonics, respectively. The bands are the systematic errors estimated from the fitting errors and the different fitting functions as discussed previously. These errors are correlated because all the errors are from the same fit parameters of the function D.

As noted above, $\bar{D}(|\Delta\eta|)$ is comprised of two parts: the contribution from the $\Delta\eta$-dependent flow fluctuation σ', and the term representing the nonflow δ. In Fig. 4.7a, these two contributors are estimated separately. The straight line is an

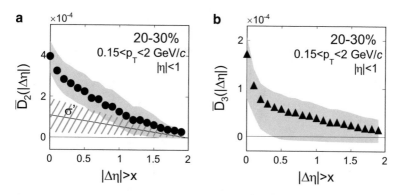

Fig. 4.7 The $\Delta\eta$-dependent component of the two-particle cumulant with $\Delta\eta$-gap, $\bar{D}$ in Eq. (4.11), of the second (**a**) and third (**b**) harmonics is shown as a function of $\Delta\eta$-gap $|\Delta\eta| > x$. (x is the x-axis value.) The *shaded bands* are systematic uncertainties. In (**a**) the estimated σ' is indicated as the *straight line*, with its uncertainty of ± 1 standard deviation as the cross-hatched area for 20–30 % central Au + Au collisions at $\sqrt{s_{NN}} = 200$ GeV. Figure reprinted from [7] under Creative Commons Attribution 4.0 License http://creativecommons.org/licenses/by/4.0/

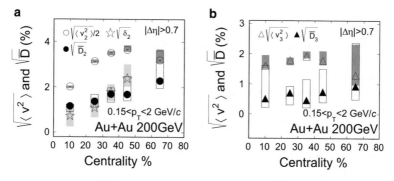

Fig. 4.8 The nonflow, $\sqrt{\bar{D}_2}$ (*solid dots*), $\sqrt{\delta_2}$ (*open stars*), $\sqrt{\bar{D}_3}$ (*solid triangles*) and flow, $\sqrt{\langle v_2^2 \rangle / 2}$ (*open circles*), $\sqrt{\langle v_3^2 \rangle}$ (*open triangles*) results are shown as a function of centrality percentile for the second (**a**) and third (**b**) harmonics, respectively. The statistical errors are smaller than the symbol sizes. The systematic errors are denoted by the vertical rectangles. Figure reprinted from [7] under Creative Commons Attribution 4.0 License http://creativecommons.org/licenses/by/4.0/

estimate of σ'. The cross-hatched area is its uncertainty of ± 1 standard deviation. The difference between the black solid points $\bar{D}(|\Delta\eta|)$ and the straight line σ' is the nonflow contribution. For both the second harmonic and the third harmonic shown in Fig. 4.7a, b, respectively, $\bar{D}(|\Delta\eta|)$ decreases as the $\Delta\eta$-gap between two particles increases.

Figure 4.8 shows $\sqrt{\langle v^2 \rangle}$ and $\sqrt{\bar{D}}$ for all measured centralities for the second harmonic (a) and the third harmonic (b) with $|\Delta\eta| > 0.7$ [5]. The errors on $\sqrt{\langle v^2 \rangle}$ and $\sqrt{\bar{D}}$ are anti-correlated. Taking $|\Delta\eta| > 0.7$, the relative magnitude $\bar{D}_2/\langle v_2^2 \rangle =$

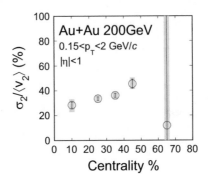

Fig. 4.9 The relative elliptic flow fluctuation $\sigma_2/\langle v_2 \rangle$ centrality dependence in $\sqrt{s_{NN}} = 200\,\text{GeV}$ Au + Au collisions. The statistical errors are shown by the *error bars*. The systematic errors are denoted by the *vertical rectangles*. Figure reprinted from [7] under Creative Commons Attribution 4.0 License http://creativecommons.org/licenses/by/4.0/

5 % $\pm$ 0.004 %(stat.) $\pm$ 2 %(sys.) for 20–30 % centrality. It is clear that $\bar{D}_2$ increases as the collisions become more peripheral.

The $\Delta\eta$-dependent nonflow contribution is primarily caused by near-side (small $\Delta\phi$) correlations. These correlations include jet correlations and resonance decays which decrease with increasing $\Delta\eta$. The $\Delta\eta$-independent correlation is dominated by anisotropic flow. However, there should be a $\Delta\eta$-independent contribution from nonflow, such as away-side dijet correlations. The away-side jet contribution should be smaller than the near-side jet contribution, because, in part, some of the away-side jets are outside the detector acceptance and, therefore, unrecorded.

Figure 4.9 shows $\sigma_2/\langle v_2 \rangle$ as a function of collision centrality classes. The $\Delta\eta$-independent relative elliptic flow fluctuation $\sigma_2/\langle v_2 \rangle$ slightly increases as collision centrality goes from more central to more peripheral. In the most peripheral centrality bin, the measurement uncertainty is large due to limited statistics.

4.3.1 Systematic Uncertainties

The systematic errors for $V\{2\}$ and $V^{1/2}\{4\}$ are estimated by varying event and track quality cuts: the primary event vertex requirement from $|z_{vtx}| < 30$ to $|z_{vtx}| < 25\,\text{cm}$; the number of fit points along the track from greater than 20–15; the distance of closest approach to the event vertex from $|dca| < 3$ to $|dca| < 2\,\text{cm}$. The systematic errors for events at 20–30 % centrality were found to be 1 % for $V_2\{2\}$ and 2 % for $V_2^{1/2}\{4\}$, and the same order of magnitude for other centralities.

The fitting error on the parameterized σ' from $\Delta V^{1/2}\{4\}$ is treated as a systematic error, which is 70 %, since σ' is consistent with zero in less than 2-σ standard deviation. Similarly, the fitting errors on the parameters used in D are treated as systematic errors that are propagated through to the total uncertainty on D. In addition, the systematic error on D from the choice of fitting function is estimated using different forms of the fitting function. The following function forms have been used in the estimation: an exponential term plus a linear term, a Gaussian function plus a linear term, an exponential function only, a Gaussian function only, and an exponential function plus a term of the form $e^{-\frac{1}{2}(\frac{\Delta\eta}{\sigma})^4}$.

The total estimated uncertainty in the second harmonic of $D(\Delta\eta)$ is an average of 40 % based on the different sources evaluated. The systematic error on D also applies to the decomposed flow through $\langle v^2 \rangle = V\{2\} - D$.

4.4 Summary

Two- and four-particle cumulant azimuthal anisotropies are analyzed between pseudo-rapidity bins in Au + Au collisions at $\sqrt{s_{NN}} = 200$ GeV from STAR. Exploiting the collision symmetry about mid-rapidity, the $\Delta\eta$-dependent and the $\Delta\eta$-independent azimuthal correlations in the data are isolated through a data-driven way. The $\Delta\eta$-independent correlation $\langle v^2 \rangle$, dominated by flow and flow fluctuations, is found to be constant over η within the measured range of ± 1 unit of pseudo-rapidity. In the 20–30 % centrality Au + Au collisions, the elliptic flow fluctuation is found to be $\sigma_2/\langle v_2 \rangle = 34 \% \pm 2 \%$(stat.) $\pm 3 \%$(sys.). The $\Delta\eta$-dependent correlation $D(\Delta\eta)$, which may be attributed to nonflow, is found to be $\bar{D}_2/\langle v_2^2 \rangle = 5 \% \pm 2 \%$(sys.) at $|\Delta\eta| > 0.7$ for $0.15 < p_T < 2$ GeV/c.

References

1. A. Bilandzic, R. Snellings, S. Voloshin, Flow analysis with cumulants: direct calculations. Phys. Rev. C **83**, 044913 (2011)
2. N. Borghini, P.M. Dinh, J.Y. Ollitrault, Is the analysis of flow at the CERN super proton synchrotron reliable?. Phys. Rev. C **62**, 034902 (2000)
3. N. Borghini, P.M. Dinh, J. Y. Ollitrault, New method for measuring azimuthal distributions in nucleus-nucleus collisions. Phys. Rev. C **63**, 054906 (2001)
4. N. Borghini, P.M. Dinh, J.Y. Ollitrault, Flow analysis from multiparticle azimuthal correlations. Phys. Rev. C **64**, 054901 (2001)
5. H. Agakishiev et al., Event-plane-dependent dihadron correlations with harmonic v_n subtraction in Au + Au collisions at $\sqrt{s_{NN}} = 200$ GeV. Phys. Rev. C **89**, 041901 (2014)
6. G. Agakishiev et al., Energy and system-size dependence of two- and four-particle v_2 measurements in heavy-ion collisions at $\sqrt{s_{NN}} = 62.4$ and 200 GeV and their implications on flow fluctuations and nonflow. Phys. Rev. C **86**, 014904 (2012)
7. N. Abdelwahab et al., Isolation of flow and nonflow correlations by two- and four-particle cumulant measurements of azimuthal harmonics in au+au collisions. Phys. Lett. B **745**, 40–47 (2015)
8. P.K. Kovtun, D.T. Son, A.O. Starinets, Viscosity in strongly interacting quantum field theories from black hole physics. Phys. Rev. Lett. **94**, 111601 (2005)

9. S. Wang, Y.Z. Jiang, Y.M. Liu, D. Keane, D. Beavis, S.Y. Chu, S.Y. Fung, M. Vient, C. Hartnack, H. Stöcker, Measurement of collective flow in heavy-ion collisions using particle-pair correlations. Phys. Rev. C **44**, 1091–1095 (1991)

10. H. Song, S.A. Bass, U. Heinz, T. Hirano, C. Shen, 200 A GeV Au+Au collisions serve a nearly perfect quark-gluon liquid. Phys. Rev. Lett. **106**, 192301 (2011)

11. A.M. Poskanzer, S.A. Voloshin, Methods for analyzing anisotropic flow in relativistic nuclear collisions. Phys. Rev. C **58**, 1671–1678 (1998)

12. L. Xu, L. Yi, D. Kikola, J. Konzer, F. Wang, W. Xie, Model-independent decomposition of flow and nonflow in relativistic heavy-ion collisions. Phys. Rev. C **86**, 024910 (2012)

13. P. Bozek, W. Broniowski, J. Moreira, Torqued fireballs in relativistic heavy-ion collisions. Phys. Rev. C **83**, 034911 (2011)

14. H. Petersen, V. Bhattacharya, S.A. Bass, C. Greiner, Longitudinal correlation of the triangular flow event plane in a hybrid approach with hadron and parton cascade initial conditions. Phys. Rev. C **84**, 054908 (2011)

15. K. Xiao, F. Liu, F. Wang, Event-plane decorrelation over pseudorapidity and its effect on azimuthal anisotropy measurements in relativistic heavy-ion collisions. Phys. Rev. C **87**, 011901 (2013)

16. J.Y. Ollitrault, A.M. Poskanzer, S.A. Voloshin, Effect of flow fluctuations and nonflow on elliptic flow methods. Phys. Rev. C **80**, 014904 (2009)

17. B.I. Abelev et al., Centrality dependence of charged hadron and strange hadron elliptic flow from $\sqrt{s_{NN}} = 200$ GeV Au + Au collisions. Phys. Rev. C **77**, 054901 (2008)

18. J. Adams et al., Azimuthal anisotropy in Au+Au collisions at $\sqrt{s_{NN}} = 200$ GeV. Phys. Rev. C **72**, 014904 (2005)

19. K. Ackermann et al., The STAR time projection chamber. Nucl. Phys. A **661**, 681–685 (1999)

20. B.I. Abelev et al.. Systematic measurements of identified particle spectra in pp, d + Au, and Au + Au collisions at the STAR detector. Phys. Rev. C **79**, 034909 (2009)

21. J. Konzer, Jet-like dihadron correlations in heavy ion collisions.   Ph.D. thesis, Purdue University, 2013

22. B. Alver et al., Non-flow correlations and elliptic flow fluctuations in gold-gold collisions at $\sqrt{s_{NN}} = 200$ GeV. Phys. Rev. C **81**, 034915 (2010)

Chapter 5
'Ridge' in d+Au

This chapter of thesis work has been summarized in [4, 5].

The deuteron–gold (d+Au) collisions at RHIC were initially proposed as a control experiment to disentangle cold nuclear effect and hot QGP final state effect for the strong suppression of high-p_T particles in central heavy-ion collisions [6–9]. The naive expectation for the small d+Au system is that only the cold nuclear matter effect needs to be considered, where little or no collective flow as a signature of QGP can develop. The observation of the long-range $\Delta\eta$ dihadron correlation at small $\Delta\phi$ (called the ridge) in $p + p$ and p+Pb collisions at the Large Hadron Collider (LHC) [10–13] was, therefore, unexpected. As discussed in Sect. 1.5, the ridge was first discovered in heavy-ion collisions and was primarily attributed to triangular anisotropic flow.

To reduce/remove contributions from jetlike correlations, low-multiplicity data were subtracted from dihadron correlations in high-multiplicity data in previous experiments. The assumptions include that there are only fundamental processes such as jet in low-multiplicity collisions, and the jetlike correlations are the same in low- and high-multiplicity collisions. The application of such a subtraction procedure led to the observation of a back-to-back ridge at $\Delta\phi \sim \pi$, along with the ridge at $\Delta\phi \sim 0$ in p+Pb collisions at $\sqrt{s_{NN}} = 5.02$ TeV. Using the same subtraction technique, the PHENIX experiment also observed a (near- and away-side) double ridge in d+Au collisions at $\sqrt{s_{NN}} = 200$ GeV within their available acceptance of $|\Delta\eta| < 0.7$ [14]. The ALICE and CMS experiment also reported multi-particle azimuthal v_n for p+Pb [15, 16]. The v_n mass ordering for the π, K, and p at low p_T, which is one of the hydrodynamic expectations, is observed in p+Pb by the ALICE experiment [17]. The PHENIX experiment also reported a v_n mass ordering for π and p in d+Au collisions [18]. Other physics mechanisms are also proposed, such as the color glass condensate, where the two-gluon density is enhanced at small $\Delta\phi$ over a wide range of $\Delta\eta$ [19–22], the initial state gluon bremsstrahlung effect calculated from the perturbative QCD and sourced by the color antennas [23], or the quantum initial anisotropy from the space momentum uncertainty principle [24].

© Springer Science+Business Media New York 2016
L. Yi, *Study of Quark Gluon Plasma By Particle Correlations in Heavy Ion Collisions*, Springer Theses, DOI 10.1007/978-1-4939-6487-1_5

If jetlike correlations are the same in high- and low-multiplicity events, the dihadron correlation difference between these two event classes would be non-jet physics. However, jet particle production contributes to the overall multiplicity. In the small collision system with relative low total multiplicity, the selection of high-multiplicity events may demand a relatively large number of jet related particles. In fact, such differences have been observed previously by STAR in the measured two-particle correlations in $p + p$ and various multiplicity d+Au collisions [3, 25]. STAR, with its large acceptance, is well suited to investigate whether and how much event selection impacts dihadron jet correlations in d+Au collisions.

5.1 Data Sample and Analysis Cuts

A total of 6.6 million Minimum-Bias (MB) d+Au collisions at $\sqrt{s_{NN}} = 200$ GeV were used in this thesis. These were triggered by the coincidence of signals from the ZDC and the BBC in the year 2003 [1]. The events were selected with the trigger setup names dAuMinBias (2.2 millions) and dAuCombined (4.4 millions) in the dataset. The data chain was reconstructed with the P04if library. The trigger ID was required to be 2001 or 2003. The reconstructed primary tracks in the TPC and the FTPCs were used in the analysis. The events' primary vertex positions in z direction were required to be within $|z_{vtx}| < 50$ cm from the TPC center. The detailed track requirements are listed in Table 5.1.

5.2 Centrality Definition in d+Au Collisions

Centrality selections in d+Au collisions are determined by three measures:

- raw charged particle multiplicity (primary tracks) within $|\eta| < 1$ in TPC,
- raw charged particle multiplicity (primary tracks) within $-3.8 < \eta < -2.8$ in FTPC Au-going side, and
- The neutral charge energy deposit in the ZDC of Au-going side quantified by the attenuated Analog-to-Digital Converter (ADC) signal.

The event requirements for each selection are listed in Table 5.2.

Table 5.1 d+Au collisions track quality cuts

TPC $|\eta| < 1$ or FTPC $2.8 < |\eta| < 3.8$
Hit points for TPC tracks ≥ 25 or FTPC tracks ≥ 5
Hit points/possible hit points ≥ 0.51
$dca < 3$ cm
$1 < p_T < 3$ GeV/c

Table 5.2 Centrality cuts

Centrality (%)	TPC	FTPC-Au	ZDC-Au
0–20	$N_{ch} \geq 29$	$N_{ch} \geq 17$	ADC ≥ 128
40–100	$N_{ch} \leq 19$	$N_{ch} \leq 9$	ADC ≤ 116

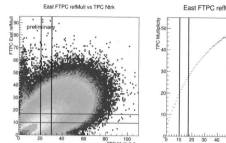
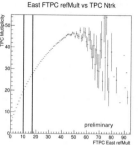
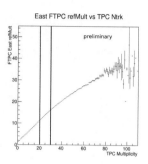

Fig. 5.1 *Left panel*: the FTPC Au-side multiplicity versus TPC multiplicity in *d*+Au collisions at $\sqrt{s_{NN}} = 200$ GeV. *Middle panel*: the mean TPC multiplicity versus FTPC-Au multiplicity. *Right panel*: the mean FTPC-Au multiplicity versus TPC multiplicity. The *color* represents the number of events, in increasing order from *blue to red*

As stated previously in Sects. 5.1 and 2.2.5, the centrality selection method can be affected by jet contributions, especially in small systems with lower multiplicities. For example, when selecting central *d*+Au collisions by demanding a high multiplicity in an event, the event in which the jets fragmented into more particles will be favored. Those jets may have large energies so that they can produce more particles, or the way they fragment biased towards more particles in the final state. In short, events with different jet fragmentations are assigned into different centrality classes.

5.2.1 Comparison Between Centrality Definitions

In order to assess the reliability of the centrality selection methods, the relationships between various centrality selection variables are studied in Figs. 5.1, 5.2, 5.3. The straight lines are the cuts for the centrality classes of 0–20 %, 20–40 %, and 40–100 %. The three left panels are the scatter plots. For example, each point in the left panel of Fig. 5.1 represents the event with certain FTPC-Au multiplicity and TPC multiplicity. The middle and the right panels show the projection of each quantity on the other quantity. For example, the middle panel in Fig. 5.1 is the average TPC multiplicity as a function of the FTPC-Au multiplicity.

In Fig. 5.1, the middle and the right panels show that there is a positive correlation between the mid-rapidity multiplicity measured by the TPC and the Au-going side forward multiplicity measured by the FTPC-Au, although the fluctuations are large as shown in the left panel.

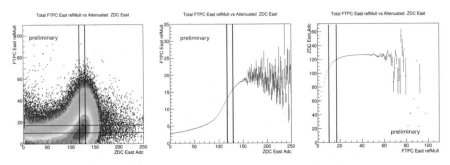

Fig. 5.2 The FTPC Au-side multiplicity versus Au-side ZDC ADC in *d*+Au collisions at $\sqrt{s_{NN}} = 200\,\text{GeV}$

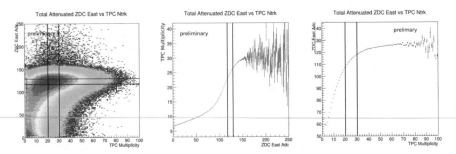

Fig. 5.3 The Au-side ZDC ADC versus TPC multiplicity in *d*+Au collisions at $\sqrt{s_{NN}} = 200\,\text{GeV}$

In Fig 5.2, the middle panel shows that the FTPC-Au multiplicity slightly increases as the ZDC-Au energy increases when the ZDC-Au ADC signal (the signal for the ZDC measured neutral energy) is less than 100. At 100 < ZDC-Au < 150, the FTPC-Au multiplicity rapidly increases with the ZDC-Au ADC. At ZDC-Au ADC > 150, the FTPC-Au multiplicity appears to saturate. The right panel shows that the ZDC energy rapidly increases as FTPC-Au multiplicity at FTPC-Au multiplicity < 10 and then it starts to saturate around FTPC-Au multiplicity > 20.

The relationship between the ZDC-Au energy and the TPC multiplicity is similar to the relationship between the ZDC-Au energy and the FTPC-Au multiplicity as shown in Fig. 5.3.

In summary, the TPC, the FTPC-Au multiplicity, and the ZDC-Au neutral energy have positive but weak correlations.

5.2.2 Comparison with Au+Au Centrality

For comparison, Figs. 5.4, 5.5, 5.6 show the relationships between the centrality estimating quantities for Au+Au collisions at $\sqrt{s_{NN}} = 200\,\text{GeV}$. They use the same data sample as Chap. 3. The TPC multiplicity and FTPC-Au multiplicity show

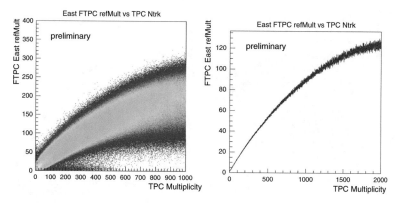

Fig. 5.4 The FTPC-Au multiplicity versus TPC multiplicity in Au+Au collisions at $\sqrt{s_{NN}}$ = 200 GeV

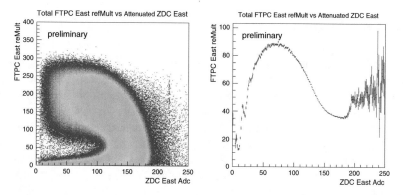

Fig. 5.5 The FTPC-Au multiplicity versus Au-side ZDC ADC in Au+Au collisions at $\sqrt{s_{NN}}$ = 200 GeV

an approximately linear relationship at low multiplicity. At high multiplicity, the FTPC-Au multiplicity starts to saturate as the TPC multiplicity increases. Figure 5.6 is similar to Fig. 2.6. While in Fig. 2.6 the mid-rapidity multiplicity is measured by the CTB, in Fig. 5.6 the mid-rapidity multiplicity is measured by the TPC. Figure 5.5 is similar to Fig. 5.6, but the FTPC-Au saturates at higher multiplicity. From Figs. 5.5 and 5.6, the multiplicity (at mid-rapidity and/or forward rapidity) and the ZDC measured neutral energy have a non-monotonic relationship. Their relationship's behavior is discussed in Sect. 2.2.3. Comparing Figs. 5.1, 5.2, 5.3 with Figs. 5.4, 5.5, 5.6, the correlation between various centrality estimating quantities in *d*+Au collisions is similar to those in peripheral Au+Au collisions (low multiplicity region) as they have similar shapes.

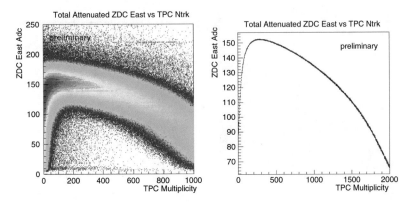

Fig. 5.6 The Au-side ZDC ADC and TPC multiplicity in Au+Au collisions at $\sqrt{s_{\mathrm{NN}}} = 200\,\mathrm{GeV}$

5.3   Correlation Analysis

The TPC-TPC dihadron correlation is measured with the trigger and associated particles both from the TPC detector. The TPC-FTPC correlation is measured with the trigger particle from the TPC and the associated particle from the FTPC. Both the correlations are normalized per trigger particle. The detector acceptance is corrected by the event-mixing technique. The dihadron correlation is given in Eq. (1.6). The details of the correlation analysis are described in Sect. 1.5.

The left panel of Fig. 5.7 shows $S(\Delta\eta, \Delta\phi)$ for the TPC-TPC correlation, and the left panel of Fig. 5.8 shows $S(\Delta\eta, \Delta\phi)$ for the TPC-FTPC correlation. Both the trigger and the associated particles are from $1 < p_T < 3\,\mathrm{GeV}/c$. The correlation data are corrected for the associated particle tracking efficiency of $85\,\% \pm 5\,\%$ (sys.) for TPC tracks, and $75\,\% \pm 5\,\%$ (sys.) for FTPC tracks [3, 25]. The efficiency does not vary from central to peripheral d+Au collisions. The right panel of Fig. 5.7 shows $B(\Delta\eta, \Delta\phi)$ for the TPC-TPC correlation, and the right panel of Fig. 5.8 shows $B(\Delta\eta, \Delta\phi)$ for the TPC-FTPC correlation. The triangle shape of the mixed event correlation is due to the η acceptance limitation of $-1 < \eta < 1$ for TPC, and $2.8 < |\eta| < 3.8$ in FTPC. The detector pair acceptance is $100\,\%$ at $\Delta\eta|_{100\,\%}$. Divided by $\langle B(\Delta\eta|_{100\,\%}, \Delta\phi)\rangle$, the mixed event correlation $B(\Delta\eta, \Delta\phi)$ is normalized to be 1 at $\Delta\eta|_{100\,\%} = 0$ for the TPC-TPC correlation, and at $\Delta\eta_{100\,\%} = \pm 3$ for the TPC-FTPC correlation (-3 for FTPC east and 3 for west). In order to describe the single particle effect in the real event, the event matching requirements are imposed to mix events with similar detector geometry and collision conditions. The mixed events are required to be within 1 cm in z_{vtx}, and to have the same multiplicity (for TPC or FTPC centrality) or within the ADC bin size of 10 (for ZDC centrality). To increase statistics, ten mixed events are performed for each trigger particle. The number of event at the large $|z_{\mathrm{vtx}}|$ is small and may have less chance to be mixed ten times. A $|z_{\mathrm{vtx}}|$ re-weight procedure is performed to make sure number of mixed events

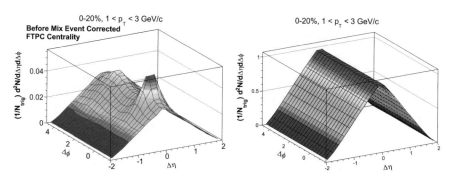

Fig. 5.7 Illustrations for real event (*left panel*) and mixed event (*right panel*) TPC-TPC correlations in 0–20 % central *d*+Au collisions with FTPC-Au multiplicity as centrality estimator

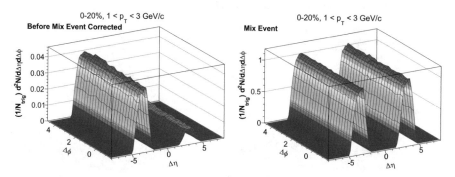

Fig. 5.8 Illustrations for real event (*left panel*) and mixed event (*right panel*) TPC-FTPC correlations in 0–20 % central *d*+Au collisions with ZDC-Au neutral energy as centrality estimator

for different $|z_{vtx}|$ have the same distribution as the real events. The two-particle correlation after mixed event corrections are shown in Figs. 5.9 and 5.10.

After the two-particle correlations are measured, two approaches are taken to analyze the data. One is to look at the associated particle correlated yield per trigger particle after ZYAM background subtraction. It is assumed that the underlying event background in *d*+Au collisions is a uniform distribution over $\Delta\phi$, with magnitude depending on $\Delta\eta$. The ZYAM background value is determined by the lowest yield of the $\Delta\phi$ distribution in each $\Delta\eta$ bin. To minimize statistical fluctuations from a single $\Delta\phi$ point, the lowest yield is calculated as the lowest average in a $\Delta\phi$ window of a certain width. The default $\Delta\phi$ width is 0.4 radians. The ZYAM systematic error is estimated by varying the $\Delta\phi$ width from 0.2 to 0.6 radians. After subtracting the yield in $C(\Delta\eta, \Delta\phi)$ by the ZYAM background, the correlated yield at its minimum will be 'zero' (in order to minimize statistical fluctuation, this zero is for the average over the chosen $\Delta\phi$ window, not for one single $\Delta\phi$ bin). After ZYAM subtraction of the underlying event, one can examine how the correlated yield varies over $\Delta\phi$ and $\Delta\eta$.

Another approach is to analyze its Fourier coefficients,

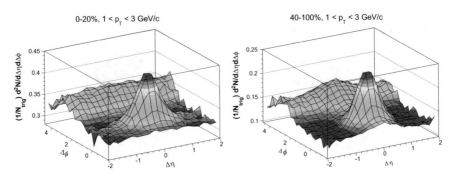

Fig. 5.9 Illustrations for the two-particle TPC-TPC correlations in 0–20 % (*left panel*) and 40–100 % (*right panel*) central d+Au collisions with FTPC-Au multiplicity as centrality estimator. The two-particle correlation is corrected by mixed events

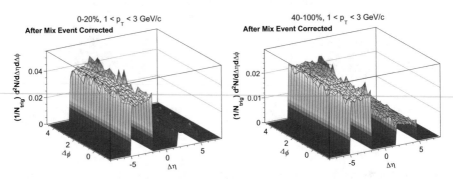

Fig. 5.10 Illustrations for the two-particle TPC-FTPC correlations in 0–20 % and 40–100 % central d+Au collisions with ZDC-Au neutral energy as centrality estimator. The two-particle correlation is corrected by mixed events

$$\frac{dN}{d\Delta\phi} \propto 1 + \sum_n 2V_n \cos(n\Delta\phi), \tag{5.1}$$

where no ZYAM subtraction procedure is needed. In Eq. (5.1), the $\Delta\phi$-independent combinatorial background goes into the first term "1". A non-zero Fourier coefficient does not necessarily mean collective anisotropic flow. Any function can be expanded in Fourier series. In order to infer possible physics from Fourier coefficients V_n, the V_n needs to be studied as a function of other observables, such as $\Delta\eta$ and multiplicity to investigate V_n behaviors. Important questions need to be asked include why and how collective flow, if any, develops in the small system and how low in multiplicity hydrodynamics still apply and QGP can still form.

5.3.1 Systematic Uncertainty

The systematic uncertainties on the correlated yields are dominated by the ZYAM background subtraction and the 5 % tracking efficiency uncertainty. They are considered in the results of the correlated yield distributions.

ZYAM is calculated as the lowest correlation magnitude averaged over a $\Delta\phi$ window of full width 0.4. The ZYAM systematic uncertainty is estimated by changing the ZYAM averaging $\Delta\phi$ width to 0.2 and to 0.6.

The tracking efficiency is 85 % for the TPC, and 75 % for the FTPC. The 5 % relative uncertainty due to the tracking efficiency correction is taken from previous publications [2, 25] (estimated by variations in the TPC gas mixture, temperature, pressure, and ionization electron drift velocity).

Systematic uncertainty in the raw correlation function is dominated by the 5 % tracking efficiency uncertainty.

5.4 Two-Particle Correlation at Mid-Rapidity

The two-particle correlation at mid-rapidity is calculated by the TPC-TPC correlation, while both the trigger and the associated particles are from TPC.

5.4.1 Central and Peripheral $\Delta\phi$ Correlations

The associated particle yields after the ZYAM subtraction are analyzed as a function of $\Delta\phi$ for three different $|\Delta\eta|$ regions: $|\Delta\eta| < 0.3$, $0.5 < |\Delta\eta| < 0.7$, and $1.2 < |\Delta\eta| < 1.8$. Both the trigger and the associated particles are taken from the TPC. The FTPC centrality selection is used instead of the TPC one for the TPC-TPC correlation in order to avoid auto-correlation from the same tracks being used for both the correlation analysis and the centrality selection.

In Fig. 5.11, the central 0–20 % collisions are represented by the red solid dots. The peripheral 40–100 % collisions data are presented by the blue solid dots. The error bars are the statistical uncertainties. The boxes are the systematic errors, which are the quadratic sums of the systematic errors due to the efficiency and the ZYAM, as well as the statistical error of the ZYAM. The ZYAM statistical error is accounted as systematic error for $\Delta\phi$ correlations because it is shared by all $\Delta\phi$ bins. The ZYAM background magnitudes are listed in the figures. The numbers in the parenthesis are the errors. The first number is the statistical error and the second two are the upper and lower systematic errors. The location of the minimum yield is indicated by the arrows.

Figure 5.11 left panel shows that, at $|\Delta\eta| < 0.3$, the near-side jet peak contributes to the near-side yield. The near-side correlated yield is two times larger than the

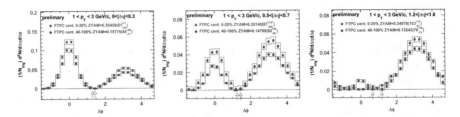

Fig. 5.11 The TPC-TPC correlation in $|\Delta\eta| < 0.3$ (*left panel*), $0.5 < |\Delta\eta| < 0.7$ (*middle panel*), and $1.2 < |\Delta\eta| < 1.8$ (*right panel*) in d+Au collisions at $\sqrt{s_{NN}} = 200\,\mathrm{GeV}$. The centrality is selected by the FTPC-Au multiplicity

away-side correlated yield for both the central and peripheral collisions. The ZYAM value is 0.3546 ± 0.0006 (stat.) $^{+0.0008}_{-0.0013}$ (sys.) for the central and 0.1578 ± 0.0005 (stat.) $^{+0.0009}_{-0.0008}$ (sys.) for the peripheral collisions. The $\Delta\phi$ locations for the ZYAM are close to each other in central and peripheral collisions, as the arrows indicate.

The $0.5 < |\Delta\eta| < 0.7$ region is the $\Delta\eta$ range used in the PHENIX [14]. The $0.5 < |\Delta\eta| < 0.7$ is still affected by the near-side jet peak, which can be seen from the large near-side peak in peripheral collisions. Figure 5.11 middle panel shows that the near-side correlated yield is slightly smaller than the away-side correlated yield for both the central and peripheral collisions. The ZYAM values are similar to but slightly smaller than those for $|\Delta\eta| < 0.3$.

The $1.2 < |\Delta\eta| < 1.8$ region is considered unaffected by the near-side jet peak, which can be seen from the zero near-side peak for peripheral collisions, while the away-side jet is still present. The ZYAM backgrounds are smaller than the mid-rapidity cases. The decease of ZYAM backgrounds from mid-rapidity to $1.2 < |\Delta\eta| < 1.8$ is larger in peripheral collisions than the decrease in central collisions.

As $|\Delta\eta|$ increases from the left panel to the right panel of Fig. 5.11, the near-side ($\Delta\phi \approx 0$) yield (the peak area) decreases, while the away-side yield ($\Delta\phi \approx \pi$) stays almost the same for both central and peripheral collisions. However, the central yield is larger than the peripheral one at both $|\Delta\phi| \approx 0$ and $|\Delta\phi| \approx \pi$ for all three $|\Delta\eta|$ windows. Even for the large $1.2 < |\Delta\eta| < 1.8$ in the right panel of Fig. 5.11, while the peripheral data are consistent with zero, the central data show a peak on the near side. Subtracting the peripheral correlation from the central one, i.e. the "central-peripheral" technique, the associated particle yields will have a double ridge structure (peaks at both $\Delta\phi = 0$ and π) for all three $|\Delta\eta|$ regions. One could attribute the difference obtained from "central-peripheral" to the ridge if one assumes that the jet correlations are the same in central and peripheral collisions and are therefore subtracted.

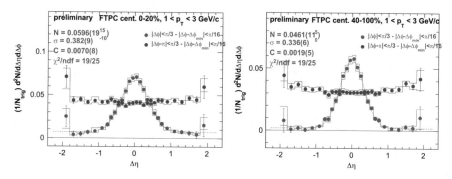

Fig. 5.12 The near-side and away-side $\Delta\eta$ projection for the FTPC-Au multiplicity selected central (*left panel*) and peripheral (*right panel*) collisions. $|\Delta\phi| < \pi/3$ is near side. $|\Delta\phi - \pi| < \pi/3$ is away side. The "$|\Delta\phi - \Delta\phi_{min}| < \pi/16$" represents the ZYAM value at minimal $\Delta\phi_{min}$ average with bin width $\pi/16$

5.4.2 The Near-Side Jetlike $\Delta\eta$ Correlations

This section quantifies the near-side jetlike correlation in central and peripheral d+Au collisions. The term "jetlike" is used instead of "jet" because other correlations are also present, such as resonance decays. The parts of the dihadron correlations used for the jet study are therefore referred to as "jetlike" correlations.

To compare the jetlike correlations in central and peripheral collisions, the near-side $\Delta\eta$ distribution is studied, because the jetlike contribution is located within a small $\Delta\eta$ angle on the near side. Figure 5.12 shows the near-side $\Delta\eta$ projection as the red symbols. The away-side yields are the blue symbols. The near side is defined as $|\Delta\phi| < \pi/3$. The away side is defined as $|\Delta\phi - \pi| < \pi/3$. The error bars are statistical uncertainties, including the statistical uncertainties from the ZYAM subtractions since the ZYAM magnitudes vary with the $\Delta\eta$. The boxes are the systematic errors from the tracking efficiency and ZYAM systematic errors. Both the magnitudes and shapes of the near-side yield are different in the central and the peripheral d+Au collisions. A Gaussian+pedestal (a single constant number) function is used to fit the near-side correlation formula: the fit results for the near-side correlations in central and peripheral collisions are listed on the plots. There are three parameters in the fit function. The Gaussian area N represents the near-side jetlike correlated yield, $Y_{jetlike}$, per radian in $\Delta\phi$. The Gaussian width σ represents the near-side jetlike peak width. The constant pedestal is C. The fit χ^2/ndf in both the central and peripheral collisions is less than 1. The near-side jetlike peak is larger and wider in central d+Au collisions than that in peripheral collisions. A similar broadening of the jetlike peak was previously observed in d+Au collisions compared with that in $p + p$ collisions [3]. The away side shapes in Fig. 5.12 are weakly dependent on $\Delta\eta$, as expected from the away-side jet or the ridge.

The "central-peripheral" $\Delta\eta$ distribution is shown in the left panel of Fig. 5.13. The near side, the red dot, has a Gaussian shape, while the away side, the blue point,

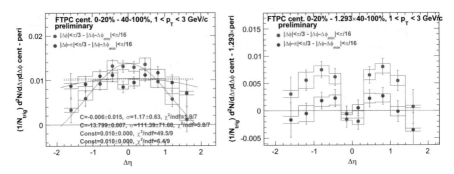

Fig. 5.13 Near-side and away-side $\Delta\eta$ projection for "central-peripheral" (*left panel*) and "central-scaled peripheral" by Eq. (5.2) (*right panel*)

is more or less constant over $\Delta\eta$. The red and blue solid curves are the Gaussian fits to the near side and the away side, respectively. The red and blue dashed curves are the constant fits to the near side and the away side. For the near side, the constant fit gives a $\chi^2/ndf = 49.5/9$ and the Gaussian fit gives a $\chi^2/ndf = 1.9/7$. For the away side, the Gaussian fit χ^2/ndf is $5.8/7$ and the constant fit χ^2/ndf is $6.4/9$. The Gaussian fit gives a very large σ, effectively consistent with the constant fit. The fit result for the "central-peripheral" is therefore consistent with a Gaussian peak on the near side and a uniform distribution on the away side. These shapes resemble jetlike features which may suggest a jetlike origin.

One way to account for the peripheral and central jet yields discrepancy is to scale the peripheral correlation up so that the peripheral near-side jetlike yield (technically, the Gaussian area on the near side) is the same as the central jetlike yield. Namely,

$$C^{\text{central}} - \alpha C^{\text{peripheral}}_{\text{jetlike}}, \tag{5.2}$$

where

$$\alpha = \frac{N^{\text{central}}_{\text{jetlike}}}{N^{\text{peripheral}}_{\text{jetlike}}}. \tag{5.3}$$

Here C^{central} and $C^{\text{peripheral}}$ are the two-particle correlations (the associated particle yields, not the fit parameter C written on the plots) in the central and peripheral collisions, respectively. $N^{\text{central}}_{\text{jetlike}}$ and $N^{\text{peripheral}}_{\text{jetlike}}$ are the fit parameters N for the central and peripheral collisions, which are the Gaussian areas representing the near-side jetlike yields. The scaling method certainly does not make the peripheral jetlike contribution look exactly the same as in the central d+Au collisions. This is because their shape difference has not been taken into account. Nevertheless, the scaling method provides a first order approximation. The scaling method with the near-side jetlike yield ratio assumes that the away-side correlated yield scales with the near-side yield from peripheral to central d+Au collisions, which is reasonable based on dijet momentum conservation.

The α parameter, the ratio of the central to peripheral near-side jetlike correlated yields, indicates how strong the event-selection effect is on the jetlike correlated yield. The ratio from the fitted yields give $\alpha = 1.29 \pm 0.05$ (stat.) ± 0.2 (sys.) for FTPC centrality 0–20 % to 40–100 %. Meanwhile, the ratio of the away-side correlated yields in central and peripheral collisions are 1.32 ± 0.02 (stat.) ± 0.01 (sys.). The α parameter for the near-side jetlike yield is consistent with the ratio of the away-side yields in central and peripheral collisions. This suggests that the away-side yield difference between central and peripheral collisions is also likely related to jets.

The result of Eq. (5.2) is shown in the right panel of Fig. 5.13. The red dots represent near side. The blue dots represent away side. After the subtraction of the scaled peripheral yield to account for the jet difference, the away-side difference between the central and peripheral collisions is consistent with zero. As aforementioned, the zero away-side yield difference suggests that the difference of the away side between central and peripheral events may primarily be due to a difference in jetlike correlations due to different event selection effect. The near-side difference is reduced. The shape of the near-side difference is the result of the subtraction of a narrow Gaussian from a wide Gaussian of equal area and an offset by a pedestal.

5.5   Event-Selection Effect on Jetlike Correlated Yield

5.5.1   Centrality Selection Methods

Besides the FTPC-Au multiplicity selection effect on the jetlike near-side correlated yield as discussed in Sect. 5.4.2, there are two other centrality selection methods: the TPC multiplicity and the ZDC-Au neutral energy. The near-side and away-side correlated yield distributions in central and peripheral collisions selected by the TPC multiplicity are shown in the left and middle panels of Fig. 5.14, while those selected by the ZDC energy ones are shown in left and middle panels Fig. 5.15. The TPC multiplicity centrality selection is expected to have a larger centrality selection effect on the jetlike correlation than the FTPC centrality one, which is observed in the right panel of Fig. 5.14. When using the TPC multiplicity for the centrality, the near-side jetlike peak is larger than the one using the FTPC multiplicity. The strong TPC centrality selection effect is due to auto-correlation (the same tracks being used for both the centrality selection and the dihadron correlation study). On the other hand, the ZDC-Au neutral energy as centrality selection has a weaker effect, as the right panel of Fig. 5.15 shows. The weaker ZDC centrality effect is expected from the large pseudo-rapidity (near the beam pipe), and the fact that the ZDC measures the neutral spectators which are not directly related to the charged particles in the

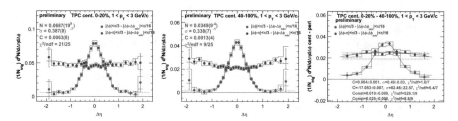

Fig. 5.14 The near-side and the away-side $\Delta\eta$ projection for the TPC multiplicity selected central (*left panel*) and peripheral (*middle panel*) collisions. The "central-peripheral" difference (*right panel*) shows a strong jetlike correlation feature

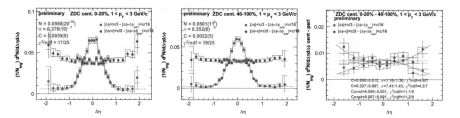

Fig. 5.15 The near-side and the away-side $\Delta\eta$ projection for the ZDC-Au energy selected central (*left panel*) and peripheral (*middle panel*) collisions. The "central-peripheral" (*right panel*) method shows weak jetlike feature

mid-rapidity used for the dihadron correlation study. Both the Gaussian and the constant fitting results are listed in the plots for the TPC and the ZDC-Au centrality selections, similar to Fig. 5.12.

5.5.2 Multiplicity Dependence

To further investigate the influence of event selection on jetlike correlations, Fig. 5.16 shows Y_{jetlike} as a function of the event activity (centrality), represented by the mid-rapidity raw (detector efficiency uncorrected) charged hadron $dN/d\eta$, in events selected according to the FTPC-Au multiplicity (solid squares) and the ZDC-Au neutral energy (open squares), respectively. The systematic uncertainties are obtained by Gaussian fits to the $\Delta\eta$ correlations varied by the ZYAM systematic uncertainties. The MB events are divided into five centrality classes as listed in Table 5.3. Figure 5.16 shows that the near-side jetlike correlated yield continues to increase with increasing event multiplicity. ALICE also reports a jetlike correlation increase with multiplicity in p+Pb collisions at $\sqrt{s_{\text{NN}}} = 5.02\,\text{TeV}$ [26]. The comparison with the HIJING model [27] is illustrated by the curve in Fig. 5.16. HIJING is a Monte Carlo program to study jet and associated particle production in high energy collisions based on QCD inspired model for multiple jets production [27]. HIJING shows no increase of jetlike yield as the multiplicity increases. The HIJING calculations are scaled down for all centrality bins by the same factor such that the lowest multiplicity bin matches real data for the trend comparison.

Table 5.3 The centrality class cuts for the FTPC multiplicity and the ZDC attenuated ADC signal

Centrality (%)	0–10	10–20	20–40	40–60	60–100
FTPC-Au multiplicity	[22,500]	[17,21]	[10,16]	[6,9]	[0,5]
ZDC-Au ADC	[133,500]	[129,132]	[117,128]	[100,116]	[0,99]

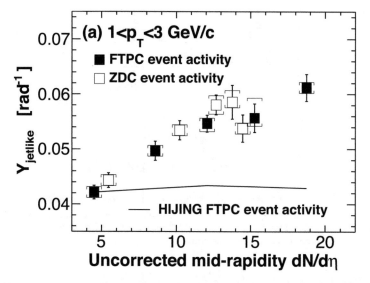

Fig. 5.16 The near-side jetlike correlated yield, obtained from a Gaussian+pedestal fit to $\Delta\eta$ distribution, as a function of the uncorrected mid-rapidity $dN/d\eta$ measured in the TPC. Two event selections are used: the FTPC-Au multiplicity (*filled squares*) and the ZDC-Au energy (*open squares*). The *curve* is the result of a HIJING calculation. *Error bars* are statistical and *caps* show the systematic uncertainties. Figure reprinted from [28] under Creative Commons Attribution 4.0 License http://creativecommons.org/licenses/by/4.0/

5.5.3 p_T Dependence

The jetlike ratio α parameter can quantify the effect of event selection on jetlike correlations. Figure 5.17 shows the p_T dependence of the α parameter. The systematic uncertainties are given by ZYAM uncertainties as in Fig. 5.16. Two sets of data points are shown. One shows the α parameter as a function of the associated particle $p_T^{(a)}$ with the trigger p_T fixed in $0.5 < p_T^{(t)} < 1\,\text{GeV}/c$. This trigger p_T range is similar to the $0.5 < p_T^{(t)} < 0.75\,\text{GeV}/c$ used by PHENIX [14]. The α parameter is larger than unity and relatively insensitive to $p_T^{(a)}$ for this particular $p_T^{(t)}$ choice. The other set of points show α as a function of trigger particle $p_T^{(t)}$ with associate particle $0.5 < p_T^{(a)} < 1\,\text{GeV}/c$ fixed. The α parameter tends to decrease with $p_T^{(t)}$.

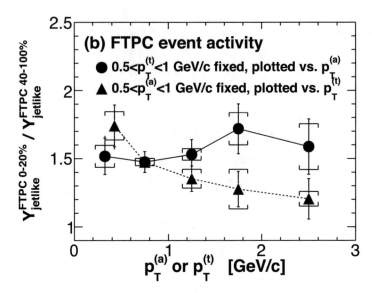

Fig. 5.17 The ratio of the correlated yields in high to low FTPC-Au multiplicity events as a function of $p_T^{(a)}$ $(p_T^{(t)})$ where $p_T^{(t)}$ $(p_T^{(a)})$ is fixed. *Error bars* are statistical and the *caps* show the systematic uncertainties. Figure reprinted from [28] under Creative Commons Attribution License 4.0 License http://creativecommons.org/licenses/by/4.0/

5.5.4 *Discussion*

There could be multiple reasons for the event-selection effect on jetlike correlations. One is a simple selection bias due to self-correlation for the centrality definition using TPC multiplicity. Such a bias may also be present for the centrality definition using FTPC-Au multiplicity: because the away-side jet can contribute to the FTPC-Au multiplicity. A high FTPC-Au multiplicity could preferentially select larger multiplicity jets (either of larger energy or happening to fragment into more particles). However, such a bias is not observed in the HIJING model implementation. Possibly because the hadrons from dijet production in HIJING for p_T range of $1 < p_T < 3\,$GeV/c may be negligible at the FTPC pseudo-rapidity region. Self-correlated centrality selection bias is unlikely present in events selected by the ZDC energy. Event centrality dependent sampling of jet energies could also be caused by physics rather than simple selection biases; for example, there could be positive correlations between jet production and the underlying event. There could also be a genuine dependence of jetlike correlations on event activity, such as initial-state k_T effects due to initial state multiple scattering or even final-state jet modifications by possible medium formation [8, 9] in the small d+Au collision system.

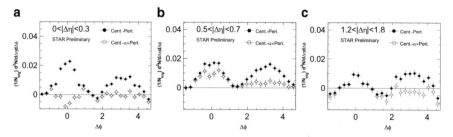

Fig. 5.18 Dihadron $\Delta\phi$ correlation difference between high- and low-multiplicity collisions in (**a**) $0 < |\Delta\eta| < 0.3$, (**b**) $0.5 < |\Delta\eta| < 0.7$ and (**c**) $1.2 < |\Delta\eta| < 1.8$ in d+Au collisions at $\sqrt{s_{NN}} = 200$ GeV for charged particles of $1 < p_T < 3$ GeV/c. Both the trigger and associated particles are from the TPC. FTPC-Au multiplicity is used for event selection. The *solid dots* represent "central-peripheral." The *open circles* represent "central-$\alpha\times$peripheral", where α is near-side Gaussian area ratio in central to peripheral collisions. The *error bars* are statistical errors. Figure reprinted from [29] with permission of ELSEVIER; permission conveyed through Copyright Clearance Center, Inc.

5.5.5 Low-Multiplicity $\Delta\phi$ Correlated Yield Subtraction

The open circles in Fig. 5.18 represent the difference between central and scaled peripheral events, with the latter scaled by the α parameter from the fit. The scaling is essentially a first order correction to the multiplicity selection effect on jetlike correlations. Indeed, the away-side yields are approximately zero for all $|\Delta\eta|$ ranges shown in Fig. 5.18. The vanish of the away-side difference suggests that the difference in the away-side long-range correlations between central and peripheral events is mostly from jetlike correlations.

The solid dots in Fig. 5.18 represent the direct difference between central and peripheral data, similar to the measurement by PHENIX [14]. The peak magnitudes on the near side and away side turn out to be similar, resembling a double ridge. As the large acceptance STAR data show, the resulting double-ridge structure may well be due to residual jetlike correlations which remain after the simple subtraction of the peripheral data from the central data.

5.6 Two-Particle $\Delta\phi$ Correlation at Forward Rapidities

The FTPCs cover $2.8 < |\eta| < 3.8$ acceptance on two sides. Studying the two-particle correlations with the trigger particle in the TPC and the associated particle in the FTPC will allow access to the large $|\Delta\eta|$ region, where the near-side jet contribution should be minimal. Figure 5.19 shows such correlations.

As Fig. 5.19 shows, on the away side, for the Au-going side, the central data (red points) are larger than the peripheral data (blue points); for the d-going side, the behavior is opposite. The difference in behaviors of the Au-going and the d-going

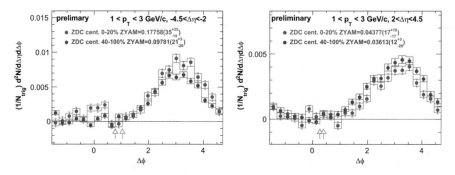

Fig. 5.19 Two-particle TPC-FTPC correlated yield $\Delta\phi$ distributions at $-4.5 < \Delta\eta < -2$ (*left panel*) and $2 < \Delta\eta < 4.5$ (*right panel*)

sides may be related to the difference in underlying parton distribution because Au-going or the d-going side correlations probe different parton x ranges (x is the momentum fraction carried by the parton). On the near side, for the Au-going side, the central yield has an excess over the peripheral yield (the latter is consistent with zero); for the d-going side, the central and peripheral yields are both consistent with zero. A recent measurement of near-side correlated yield in p+Pb collisions at forward rapidities by LHCb collaboration is reported at [30]. They found the near-side correlated yield is more pronounced in Pb-going side than p-going side. However, their correlated yields are compatible when comparing events with similar absolute local multiplicity.

5.7 Near-Side Long-Range Ridge $\Delta\eta$ Dependence

To further understand the near-side ridge, the near-side and away-side correlated yields in central d+Au collisions are plotted as a function of $\Delta\eta$ with both the TPC-TPC and the TPC-FTPC correlations on the same graph in Fig. 5.20. To avoid self-correlations, the ZDC-Au centrality selection is used for both correlations. For comparison, the ZYAM magnitude is also plotted. The points at $\Delta\eta < -2$ and the points at $\Delta\eta > 2$ are from the TPC-FTPC correlations. The others are from the TPC-TPC correlations. The near-side yields are the pink circles. The away-side yields are the blue triangles. The ZYAM values are represented by the red lines. There is a jump/drop at $|\Delta\eta| = 2$ between the two sets of correlation data. This discontinuity is due to the fact that the particle pairs with $\Delta\eta$ to the left of $|\Delta\eta| = 2$ and those to the right come from different kinematic regions (TPC particles have $-1 < \eta < 1$ and FTPC particles have $-3.8 < \eta < -2.8$ or $2.8 < \eta < 3.8$) even though their $\Delta\eta$ gaps are similar at the step. If there is detector covering $1 < |\eta| < 2.8$, the distribution will be continuous.

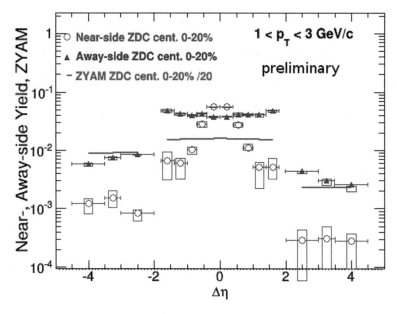

Fig. 5.20 The $\Delta\eta$ dependence of the near-side and away-side correlated yields and the estimated ZYAM background (scaled by 1/20)

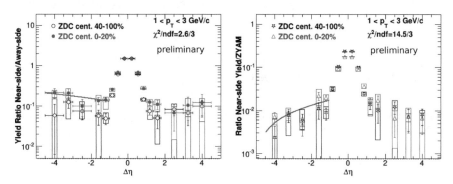

Fig. 5.21 *Left panel*: the $\Delta\eta$ dependence the ratio of the near-side to away-side correlated yields. The *solid line* is a linear fit to the ratio in central d+Au collisions, yielding a slope of $(-2.2 \pm 1.8) \times 10^{-2}$. *Right panel*: the $\Delta\eta$ dependences the ratio of the near-side correlated yield to the ZYAM values. The *solid line* is a linear fit to the ratio in central d+Au collisions, yielding a slope of $(4 \pm 1) \times 10^{-3}$

To possibly elucidate the formation mechanism of the ridge, the ratio of the near-side to the away-side correlated yields is studied in left panel of Fig. 5.21. The solid dots are the central data. The open circles are the peripheral data. While the large peak at $\Delta\eta \approx 0$ is due to the near-side jet, the ratio is rather uniform in $\Delta\eta$ at $|\Delta\eta| > 1$. A linear fit to the $\Delta\eta < -1$ region gives a slope of $(-2.2 \pm 1.8) \times 10^{-2}$. The linear fit indicates that the ratio is consistent with a constant within the standard deviation.

Since the away-side correlated yields are dominated by jets, the $\Delta\eta$-independent ratio may hint a connection between the near-side ridge and dijet production, even though any possible jet contribution to the ridge at $|\Delta\eta| > 1$ should be minimal.

However, the near-side ridge does not seem to scale with the ZYAM magnitude. A linear fit to the ratio of the near-side correlated yield over ZYAM indicates a slope of $(4 \pm 1) \times 10^{-3}$ in $\Delta\eta$ which differs from zero by four times the standard deviations, as the right panel of Fig. 5.21 shows.

5.8   Fourier Coefficients

The above correlated yields are subject to ZYAM background subtraction. Another way to study the correlations is to use Fourier series to characterize azimuthal functions. The Fourier coefficients are calculated by

$$V_n(\Delta\eta) = \langle \cos(n\Delta\phi) \rangle = \frac{\int_0^{2\pi} C(\Delta\eta, \Delta\phi) \cos(n\Delta\phi) d\Delta\phi}{\int_0^{2\pi} C(\Delta\eta, \Delta\phi) d\Delta\phi}. \tag{5.4}$$

$C(\Delta\eta, \Delta\phi)$ is the correlation function, see Eq. (1.6) in Sect. 5.3. V_n is the average of the $\cos(n\Delta\eta)$ over the trigger-associated pairs in the selected $\Delta\eta$ window for all events.

5.8.1   Systematic Uncertainty

Systematic uncertainties in the Fourier coefficients in Fig. 5.25 are estimated to be 5 % for V_1 and 10 % for V_2, while V_3 is consistent with zero within 2σ. It is estimated by varying the *dca* from the default 3 to 2 cm, and varying number of hit points from the default 25 points to 20 points (see Figs. 5.22, 5.23, 5.24).

5.8.2   Results

The left panel of Fig. 5.25 shows the Fourier coefficient V_1, the middle panel shows the V_2 and the right panel shows the V_3. V_3 is mostly consistent with zero. Three $\Delta\eta$ ranges for the correlations are shown: the TPC-FTPC Au-side, the TPC-TPC, and the TPC-FTPC d-side correlations. Results from all three centrality definitions are shown, plotted at the corresponding measured mid-rapidity charged particle $dN/d\eta$. V_2 is finite at all measured $\Delta\eta$, and is larger at mid-rapidity than at forward/backward rapidities; V_2 from the TPC-FTPC d-side correlation may be even larger than that from the TPC-FTPC Au-side correlation. As shown in the

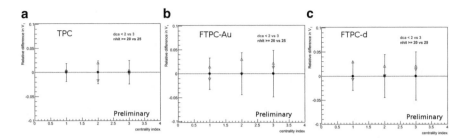

Fig. 5.22 Systematic error estimation (relative error $\frac{V_1 - V_1^{\text{default}}}{V_1^{\text{default}}}$) for Fourier coefficient V_1 by changing track cuts for TPC-TPC ($1.2 < |\Delta\eta| < 1.8$), TPC-FTPC Au-side ($-4.5 < \Delta\eta < -2$), and TPC-FTPC d-side ($2 < \Delta\eta < 4.5$) correlations with FTPC Au-side multiplicity event selection. (**a**) TPC. (**b**) FTPC-Au. (**c**) FTPC-d

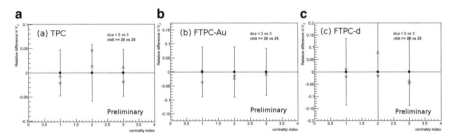

Fig. 5.23 Systematic error estimation (relative error $\frac{V_2 - V_2^{\text{default}}}{V_2^{\text{default}}}$) for Fourier coefficient V_2 by changing track cuts for TPC-TPC ($1.2 < |\Delta\eta| < 1.8$), TPC-FTPC Au-side ($-4.5 < \Delta\eta < -2$), and TPC-FTPC d-side ($2 < \Delta\eta < 4.5$) correlations with FTPC Au-side multiplicity event selection. (**a**) TPC. (**b**) FTPC-Au. (**c**) FTPC-d

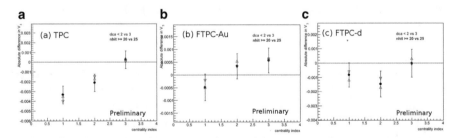

Fig. 5.24 Systematic error estimation (absolute error $V_3 - V_3^{\text{default}}$) for Fourier coefficient V_3 by changing track cuts for TPC-TPC $1.2 < |\Delta\eta| < 1.8$, TPC-FTPC Au-side $-4.5 < \Delta\eta < -2$, and TPC-FTPC d-side $2 < \Delta\eta < 4.5$ with FTPC Au-side multiplicity event selection. (**a**) TPC. (**b**) FTPC-Au. (**c**) FTPC-d

left panel of Fig. 5.25, V_1 varies approximately as $(dN/d\eta)^{-1}$, and from the middle panel of Fig. 5.25, V_2 is approximately independent of $dN/d\eta$. As a result, the "central-peripheral" correlated yield (essentially products of multiplicity and V_n) could be dominated by a V_2 component, with a magnitude similar to those for the

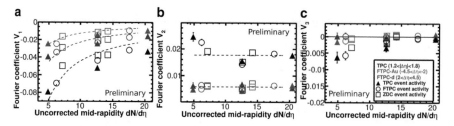

Fig. 5.25 Fourier coefficients V_1 (*left panel*), (**a**), V_2 (*middle panel*), (**b**) and V_3 (*right panel*), (**c**) versus the measured mid-rapidity charged particle density $dN/d\eta$

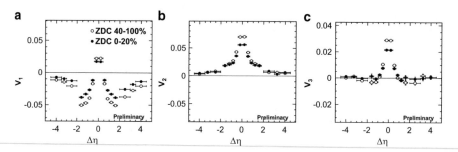

Fig. 5.26 Fourier coefficient V_1 (*left panel*), (**a**) V_2 (*middle panel*), (**b**) and V_3 (*right panel*), (**c**) versus $\Delta\eta$

individual peripheral and central data. There appears to be a symmetric back-to-back double ridge at large $\Delta\eta$ in "central-peripheral" correlations. After accounting for multiplicity biases, the "central-scaled-peripheral" correlated yield is essentially eliminated on the away side, as shown by the open symbols in Fig. 5.18.

Figure 5.26 shows the second harmonic Fourier coefficient V_n as a function of $\Delta\eta$ for both central and peripheral collisions. The near-side jet peak at $\Delta\eta = 0$ contributes to all V_n. V_2 continues to decrease with increasing $|\Delta\eta|$ from the small $|\Delta\eta|$ jet region to the large $|\Delta\eta|$ ridge region. The V_2 values are similar between central and peripheral d+Au collisions. A V_2 from the jet correlation decreases as $|\Delta\eta|$ increases. However, if V_2 at large $\Delta\eta$ is of a hydrodynamic collective flow origin, the decreasing trend with $\Delta\eta$ is not unreasonable. However, the similar V_2 magnitudes in central and peripheral collisions seem surprising in the hydrodynamic collective flow picture.

5.9 Summary

Dihadron correlations are measured at mid-rapidity and forward/backward rapidities using the STAR TPC and FTPC as a function of centrality, i.e. event activity, in d+Au collisions at $\sqrt{s_{NN}} = 200$ GeV. The centrality is classified by three measurements: the mid-rapidity TPC charged particle multiplicity, the FTPC-Au forward charged particle multiplicity, and the ZDC-Au zero-degree neutral energy.

The correlated yields are extracted by subtracting the estimated $\Delta\phi$-independent combinatorial background using the ZYAM method. It is found that the correlated yield is larger in central than peripheral collisions, and that the $\Delta\eta$-dependence of the observed yield difference resembles jetlike features, suggesting a possible jetlike origin. There could be multiple reasons for this difference, ranging from simple auto-correlation biases to physical differences between central and peripheral d+Au collisions. After scaling the peripheral data by the ratio of the near-side jetlike correlated yields, the away-side correlation difference is significantly diminished. This analysis demonstrates that the long-range dihadron correlation difference between central and peripheral events at RHIC may primarily be due to jets. Such event-selection effects on jetlike correlations must be addressed before investigating possible non-jet correlations, such as anisotropic flow, in d+Au collisions at RHIC.

Finite near-side correlated yields are present above the estimated ZYAM background in central d+Au collisions at large $\Delta\eta$ between particle pairs both from the TPC, and one particle from the TPC and the other from the Au-beam direction FTPC. The near-side ridge at $|\Delta\eta| > 1$ appears to scale with the away-side correlated yield, which is believed that the away-side yield is dominated by jet contributions in d+Au collisions at $\sqrt{s_{NN}} = 200$ GeV. The near-side ridge does not scale with the estimated ZYAM background.

Fourier coefficients of the raw dihadron correlations are also reported. All $\Delta\phi$ correlation functions appear to have a V_1 and a V_2 Fourier component. The V_1 is found to be approximately inversely proportional to event multiplicity. The V_2 is found to decrease with $\Delta\eta$, but remains finite at both forward and backward rapidities of $|\Delta\eta| \approx 3$ with similar magnitudes. The V_2 is found approximately independent of the event multiplicity. Extreme caution should be taken when interpreting the V_2 result in peripheral collisions as being from jets and in central d+Au collisions as primarily being from non-jet, elliptic flow physics.

References

1. K. Ackermann et al., STAR detector overview. Nucl. Instrum. Methods **A499**(2–3), 624–632 (2003)
2. G. Agakishiev et al., Energy and system-size dependence of two- and four-particle v_2 measurements in heavy-ion collisions at $\sqrt{s_{NN}} = 62.4$ and 200 GeV and their implications on flow fluctuations and nonflow. Phys. Rev. C **86**, 014904 (2012)
3. B.I. Abelev et al.. Systematic measurements of identified particle spectra in pp, $d + $ Au, and Au $+$ Au collisions at the STAR detector. Phys. Rev. C **79**, 034909 (2009)
4. L. Adamczyk et al., Effect of event selection on jetlike correlation measurement in d+Au collisions at $\sqrt{s_{NN}} = 200$ GeV. Phys. Lett. B **743**, 333–339 (2015)
5. L. Adamczyk et al., Long-range pseudorapidity dihadron correlations in collisions at d+Au collisions at $\sqrt{s_{NN}} = 200$ GeV. Phys. Lett. B **747**, 265–271 (2015)
6. I. Arsene et al., Quark-gluon plasma and color glass condensate at RHIC? the perspective from the BRAHMS experiment. Nucl. Phys. A **757**(1–2), 1–27 (2005)
7. B. Back et al., The PHOBOS perspective on discoveries at RHIC. Nucl. Phys. A **757**(1–2), 28–101 (2005)
8. J. Adams et al., Experimental and theoretical challenges in the search for the quark-gluon plasma: the STAR Collaboration's critical assessment of the evidence from RHIC collisions. Nucl. Phys. A **757**(1–2), 102–183 (2005); First Three Years of Operation of RHIC

9. K. Adcox et al., Formation of dense partonic matter in relativistic nucleus-nucleus collisions at RHIC: experimental evaluation by the PHENIX collaboration Nucl. Phys. A **757**(1–2), 184–283 (2005); First Three Years of Operation of RHIC

10. S. Chatrchyan et al., Observation of long-range, near-side angular correlations in pPb collisions at the LHC. Phys. Lett. B **718**(3), 795–814, (2013)

11. B. Abelev et al., Long-range angular correlations on the near and away side in p-Pb collisions at $\sqrt{s_{NN}} = 5.02$ TeV. Phys. Lett. B **719**(1–3), 29–41 (2013)

12. G. Aad et al., Observation of associated near-side and away-side long-range correlations in $\sqrt{s_{NN}} = 5.02$ TeV proton-lead collisions with the ATLAS detector. Phys. Rev. Lett. **110**, 182302 (2013)

13. G. Aad et al., Measurement of long-range pseudorapidity correlations and azimuthal harmonics in $\sqrt{s_{NN}} = 5.02$ TeV proton-lead collisions with the ATLAS detector. Phys. Rev. C **90**, 044906 (2014)

14. A. Adare et al., Quadrupole anisotropy in dihadron azimuthal correlations in central d+Au collisions at $\sqrt{s_{NN}} = 200$ GeV. Phys. Rev. Lett. **111**, 212301 (2013)

15. B. Abelev et al., Multiparticle azimuthal correlations in p-Pb and Pb-Pb collisions at the CERN large hadron collider. Phys. Rev. C **90**, 054901 (2014)

16. V. Khachatryan et al., Evidence for collective multiparticle correlations in p-Pb collisions. Phys. Rev. Lett. **115**, 012301 (2015)

17. B.B. Abelev et al., Long-range angular correlations of π, K and p in p-Pb collisions at $\sqrt{s_{NN}} = 5.02$ TeV. Phys. Lett. B **726**, 164–177 (2013)

18. A. Adare et al., Measurement of long-range angular correlation and quadrupole anisotropy of pions and (anti)protons in central d + Au collisions at $\sqrt{s_{NN}} = 200$ GeV. Phys. Rev. Lett. **114**, 192301 (2015)

19. A. Dumitru, K. Dusling, F. Gelis, J. Jalilian-Marian, T. Lappi, R. Venugopalan, The ridge in proton-proton collisions at the LHC. Phys. Lett. B **697**(1) 21–25, (2011)

20. A. Dumitru, J. Jalilian-Marian, E. Petreska, Two-gluon correlations and initial conditions for small x evolution. Phys. Rev. D **84**, 014018 (2011)

21. P. Tribedy, R. Venugopalan, QCD saturation at the LHC: comparisons of models to p+p and A+A data and predictions for p+Pb collisions. Phys. Lett. B **710**(1), 125–133 (2012)

22. P. Tribedy, R. Venugopalan, Erratum to "QCD saturation at the LHC: comparisons of models to and data and predictions for collisions" [Physics Letters B 710 (1) (2012) 125] . Phys. Lett. B **718**(3), 1154 (2013)

23. M. Gyulassy, P. Levai, I. Vitev, T. Biró, Initial-state bremsstrahlung versus final-state hydrodynamic sources of azimuthal harmonics in at RHIC and LHC. Nucl. Phys. A **931**, 943–948 (2014)

24. D. Molnar, F. Wang, C.H. Greene, Momentum anisotropy in nuclear collisions from quantum mechanics. arXiv:1404.4119 (2014)

25. J. Adams et al., Evidence from d + Au measurements for final-state suppression of high-p_T hadrons in Au + Au collisions at RHIC. Phys. Rev. Lett. **91**, 072304 (2003)

26. B. Abelev et al., Multiplicity dependence of jet-like two-particle correlation structures in p-Pb collisions at $\sqrt{s_{NN}} = 5.02$ TeV. Phys. Lett. B **741**, 38–50 (2015)

27. M. Gyulassy, X.N. Wang, HIJING 1.0: a monte carlo program for parton and particle production in high energy hadronic and nuclear collisions. Comput. Phys. Commun. **83**(2–3), 307–331 (1994)

28. L. Adamczyk et al., Effect of event selection on jetlike correlation measurement in d+au collisions at $\sqrt{s_{NN}} = 200$ gev. Phys. Lett. B **743**, 333–339 (2015)

29. L. Yi, Search for ridge in d+au collisions at RHIC by STAR. Nucl. Phys. A **931**, 326–330 (2014); Quark Matter 2014XXIV International Conference on Ultrarelativistic Nucleus-nucleus Collisions.

30. R. Aaij et al., Measurements of long-range near-side angular correlations in $\sqrt{s_{NN}} = 5$ TeV proton-lead collisions in the forward region. arXiv 1512.00439 (2015)

Chapter 6
Conclusion

Three distinct but intellectually connected measurements are reported in this thesis. The first measurement regards triangular harmonic flow in heavy-ion collisions, which is considered to be the primary source for the novel near-side ridge and away-side double-peak in two-particle angular correlations. The second one regards isolation of flow and nonflow effects in two-particle correlations, which is critical in extracting QGP properties, such as the shear viscosity to entropy density ratio η/s. The third one regards long-range ridge correlations in d+Au collisions, which have important implications to possible QGP formation in small systems.

The ridge is a particle correlation with small azimuthal opening-angle but long-range in pseudo-rapidity separation. The double-peak correlation refers to two peaks in the jet recoil direction in azimuth with respect to the triggered one. Both were observed in heavy-ion collisions after the subtraction of elliptic flow background, and can be quantified by triangular flow v_3. The $\Delta\eta$-gap, centrality and p_T dependence of v_3 is measured in Au+Au collisions at $\sqrt{s_{NN}} = 200\,\text{GeV}$. The hydrodynamic calculation with a lumpy initial condition describes well the measured v_3 below $p_T < 2\,\text{GeV}/c$. The measurement helps constrain the η/s parameter in hydrodynamic model calculations.

The isolation of flow (a global correlation) and nonflow (few-body correlations) exploits the measurements of two- and four-particle azimuthal cumulants in symmetric Au+Au collisions. A data-driven method is applied to separate the $\Delta\eta$-dependent and $\Delta\eta$-independent azimuthal correlations. The $\Delta\eta$-independent correlation is dominated by flow and flow fluctuations. It is found to be constant over η in the measured range $|\eta| < 1$. The relative flow fluctuation is found be to $34\% \pm 2\%(\text{stat.}) \pm 3\%(\text{sys.})$ in 20–30 % central Au+Au collisions at $\sqrt{s_{NN}} = 200\,\text{GeV}$. The $\Delta\eta$-dependent correlation may be attributed to nonflow. It is found to be $5\% \pm 2\%$ relative to the square of the average flow with *eta*-gap $|\Delta\eta| > 0.7$ and particle $0.15 < p_T < 2\,\text{GeV}/c$ in 20–30 % central Au+Au collisions at $\sqrt{s_{NN}} = 200\,\text{GeV}$.

© Springer Science+Business Media New York 2016

L. Yi, *Study of Quark Gluon Plasma By Particle Correlations in Heavy Ion Collisions*, Springer Theses, DOI 10.1007/978-1-4939-6487-1_6

While Au+Au collisions have large underlying event with collective flow, particle production in d+Au collisions is dominated by jet fragmentation process at modest p_T. It is found that the jetlike correlations in d+Au collisions depend on the event activity, in contrast to initial expectations. To account for the event activity dependence of jetlike correlations, a scaling factor is applied to the low-activity data before it is subtracted from the high-activity data. The remaining nonjet correlation is minimal on the away side. On the near side, a finite correlated yield is observed to extend to large pseudo-rapidity distances in high-activity collisions. This so-called near-side ridge appears to scale with the away-side jet as a function of $\Delta\eta$. A Fourier analysis of the measured azimuthal correlations indicates a V_2, independent of collision activity, with similar magnitudes between Au- and d-going directions. These measurements help constrain theoretical models for the ridge in d+Au collisions.

Appendix
Kinematic Variables

When dealing with relativistic heavy-ion collisions, several kinematic variables are defined in the way that they have simple Lorentz transformation forms. The natural units are used, $c = \hbar = 1$, where c is the speed of light and $\hbar$ is the Planck constant.

The contravariant vector of a particle with momentum $\vec{p}$ and energy E is

$$p^{\mu} = (E, \vec{p}) = (E, \vec{p}_T, p_z) = (E, p_x, p_y, p_z), \qquad (A.1)$$

where z is the beam direction and x–y is the transverse plane. The transverse momentum magnitude is $p_T = |\vec{p}_T| = \sqrt{p_x^2 + p_y^2}$, and the azimuthal angle ϕ spans from the x-axis to the $\vec{p}_T$ vector.

The rapidity of a particle is defined as

$$y = \frac{1}{2} \ln \frac{E + p_z}{E - p_z}. \qquad (A.2)$$

Rapidity is a dimensionless variable. The advantage of using rapidity is that its Lorentz transformation follows simple additive law. For example, a particle has rapidity y in one frame, and rapidity y' in another frame moving at a velocity β in the z-direction relative to the first. The Lorentz transformation of the particle rapidity is simply given by

$$y' = y - y_{\beta} \qquad (A.3)$$

where

$$y_{\beta} = \frac{1}{2} \ln \frac{1 + \beta}{1 - \beta}. \qquad (A.4)$$

© Springer Science+Business Media New York 2016
L. Yi, *Study of Quark Gluon Plasma By Particle Correlations in Heavy Ion Collisions*, Springer Theses, DOI 10.1007/978-1-4939-6487-1

Following equations relate rapidity y and other kinematic variables:

$$E = m_T \cosh y, \tag{A.5}$$

$$p_z = m_T \sinh y. \tag{A.6}$$

where transverse mass $m_T = \sqrt{m^2 + p_T^2}$.

Despite the convenience of rapidity, the particle mass (or particle species) is not easy to measure in heavy-ion experiments. The pseudo-rapidity η is therefore often used as a substitute for y. The η is given by the particle momentum polar angle θ relative to the beam axis:

$$\eta = -\ln \tan \frac{\theta}{2} = \frac{1}{2} \ln \frac{|\vec{p}| + p_z}{|\vec{p}| - p_z}. \tag{A.7}$$

For massless particles, $\eta = y$. For mid-rapidity particles, $\beta_z \ll 1$: $\eta \approx y$. Similarly the following relations are useful:

$$|\vec{p}| = p_T \cosh \eta, \tag{A.8}$$

$$p_z = p_T \sinh \eta. \tag{A.9}$$

VITA

Li Yi graduated from the University of Science and Technology of China in July 2010 with a Bachelor of Science Degree in Physics. She pursued her graduate study in Purdue University from 2010 to 2014. Li joined STAR experiment in the Relativistic Heavy Ion Collider, Brookhaven National Lab, Upton, NY at year 2010. Li worked on the collective flow and jet correlation measurements by studying the particle correlations in d+Au and Au+Au collisions.

© Springer Science+Business Media New York 2016   83
L. Yi, *Study of Quark Gluon Plasma By Particle Correlations in Heavy Ion Collisions*, Springer Theses, DOI 10.1007/978-1-4939-6487-1